Gender and Innovation in the New Economy

Seppo Poutanen • Anne Kovalainen

Gender and Innovation in the New Economy

Women, Identity, and Creative Work

Seppo Poutanen
Turku School of Economics
University of Turku
Turku, Finland

Anne Kovalainen
Turku School of Economics
University of Turku
Turku, Finland

ISBN 978-1-137-52700-4   ISBN 978-1-137-52702-8 (eBook)
DOI 10.1057/978-1-137-52702-8

Library of Congress Control Number: 2017936742

Cover image © Ruben Tresserras / Getty Images

Printed on acid-free paper

This Palgrave Macmillan imprint is published by Springer Nature
The registered company is Nature America Inc.
The registered company address is: 1 New York Plaza, New York, NY 10004, U.S.A.

Acknowledgements

The idea for this book was born during our stay as visiting fellows at Stanford University, at the Clayman Institute for Gender Research and School of Humanities and Sciences. We are grateful for the invitation to join those stimulating and lively surroundings and for the continued support of our many colleagues there, Director of the Clayman Institute, Professor Shelley Correll, Executive Director Lori Nishiura Mackenzie, Professor Helen E. Longino and Professor Valerie Miner. Warm thanks to you all!

We thank our colleagues at Turku School of Economics for a stimulating and friendly atmosphere. Entrepreneurship Unit is a good and inspiring place to work.

Invaluable assistance in the completion of this book has been provided by our research assistant Matti W. Karinen.

Our thanks also go to those who commissioned and assisted in the production of this work at Palgrave Macmillan, particularly Marcus Ballenger and his superb team. Thank you to Jazmine Robles who most efficiently took care of practicalities in the publication process.

To end, the authors wish to thank the Academy of Finland for the funding and grants that supported this work. In particular, for the Minna Canth Academy Professorship (218207, 263829) and Strategic Research Council at the Academy of Finland for strategic research consortium funding for 'Smart Work in Platform Economy' (www.smartwork research.fi) (303667).

CONTENTS

1 Setting the Scene   1
   References   7

2 Gender in Inventions and Innovations   9
   2.1 Gendering Early Inventions   10
   2.2 Gendered Patterns of Patenting   18
   2.3 Science Changing Patterns?   23
   2.4 Widening the Field of Innovations   32
   References   38

3 New Economy, Platform Economy and Gender   47
   3.1 What is the New Economy?   48
   3.2 Changing Relationship Between Gender, Work and Capital   58
   3.3 Platform Economy, Gig Economy and Sharing Economy   73
   3.4 Gendering the Platform Economy   83
   References   86

4 Innovations, Gender and the New Economy   97
   4.1 Gender and Gaming Industries   98
   4.2 Gender and Technical Design   108
   4.3 Girls Changing Codes?   115
   4.4 Gendering the Internet of Things   119
   References   127

5 Creative Work and Gender 135
5.1 Spanning the Boundaries of Creative Work 135
5.2 Care and Technological Innovations 147
5.3 Hybridization of Care Work 157
References 162

6 Envisioning the Future 169
6.1 Gender and Innovations: Turning the Tide 169
6.2 Gender and Innovations: Widening the Field 172
6.3 Concluding Remarks 178
References 182

Index 185

CHAPTER 1

Setting the Scene

Imagine modern society without innovation. Your soya latte would not be part of your breakfast. The electric light would not have replaced the gaslight in your kitchen and you would not take the bus to work, nor would you be able to read the news or listen to music on your mobile phone while you sit in the bus. Neither would you be able to make calls on the move. Indeed, these innovations have become part of our everyday lives to the extent that we no longer pay their existence much attention. Thinking about contemporary everyday life in a modern society without these innovations is impossible. Our lives are entangled with them. Major power cuts, the havoc caused by failures in design and natural disasters show us how complex our ways of life are, and how dependent we are on interconnected innovations built in relation to one another.

Innovations transform and continuously mould everyday lives and have become part of the value of industrial material products as well as imma-terial services. This happens most often gradually, but sometimes radically. Some inventions have a global impact: electricity and many technical devices have laid the technological foundations that have allowed product and service innovations to further develop and refine all sorts of products that enhance our lives today.

This book deals with innovation and gender. It explores women's inven-tions and innovations and recognition of that work, including the creative work leading to innovations and the varied forms of innovation, ranging from social to technological innovations. The contexts of innovations are crucial

© The Author(s) 2017
S. Poutanen, A. Kovalainen, *Gender and Innovation in the New Economy*, DOI 10.1057/978-1-137-52702-8_1

because innovations seldom occur in isolation or as the result of work done by one person alone. In this book, the new and the old economy are analysed in relation to innovations. In the following sections we will also focus on examples which illustrate inventions and innovations by women and show some interesting examples affecting everyday life. Equally, it is also essential to realize, that there is no reason for discussion about women to be held distinct from discussion related to innovation. This is not only because many innovations occur because of female innovators, but also because innovations are often designed specifically for women, for men, for specific age groups or for other certain types of users. More generally, the incremental process of innovation has elements of gender involved in many ways, even if we do not notice it as consumers. Gender-related elements are considered and attached to many products or services, even before they appear (Poutanen and Kovalainen 2013, 2016). This is in marked contrast to efforts to argue the contrary by attempting to separate gender from innovation (e.g. Bath 2014).

How then can we explain the heavily gendered nature of innovation? This book will offer some systematic thinking and answers to the question of how and in what ways innovations are gendered.

There is extensive research on the gender imbalance in research and academia, ranging from gender segregation of education fields to leaky pipelines in academic careers (e.g. Schiebinger and Klinge 2013). There is also extensive literature on gender differences and similarities in inventions, patenting and innovations, addressing the question from various points of view: these range from the share of men and women in patenting, to gender related aspects of innovations, participation in research and development work, and the gender gap in start-ups and venture capital firms (e.g. Bunker-Whittington and Smith-Doerr 2005, 2008; Rosser 2009, 2012; Lins and Lutz 2016). Fewer examples in recent and novel research can be found on the reasons for and effects of the gender imbalance in research-based innovations and in their spread and use. Through knowledge and technology transfer, the results of research are transferred to new products, new processes and new practices in both the private and public sectors as well as in civil society (Best et al. 2016; Etzkowitz et al. 2000; Hellström and Jacobs 2005; McKelvey and Holmèn 2009; Abreu and Grinevitch 2013, 2017).

Gender disparities – the differences in the numbers of women and men – in science and engineering professions are enduring topics in both science and innovation policy analyses and scholarly debates. The gender differences in professions and in career positions relating to and leading towards invention and innovative activity are well reported. The underrepresentation of women in the fields of technology, innovation and entrepreneurship, as well as in science commercialization (e.g. Rosser 2009; Bunker-Whittington and Smith-Doerr 2008; Wacjman 2004) are all well known facts. However, the current situation – after the global financial crisis of 2008–2010, which has led to a major decline in R&D&I in many countries – is seldom understood to be an ecosystem issue. What would happen if gender disparity in degree education in sciences and engineering ceased to be as prevalent as it is now? According to estimates by Hunt and her colleagues, closing the gender gap in science and especially among engineering degree holders in the USA would increase US GDP per capita by 2.7% (Hunt et al. 2013). This is important because education is part of the innovation ecosystem, and as the route to an innovation society, education has an important role in innovation and invention outcomes. The ecosystem as a concept is to some extent problematic, as it tends to smooth away many of the 'wrinkles' – the fractures and the dissonances in the ways organizations in the ecosystem and within culture function – and thus it masks the deeply gendered nature of the ecosystem thinking. Ecosystem emphasizes the high interdependency of activities and institutions.

Taking a specifically economic and human capital perspective of gender, it can be argued that women are 'underutilized' as a human resource in science and its commercialization. From the science education point of view, the scarcity of female participation in innovation and in commercial knowledge transfer (e.g. Link et al. 2007) adds an entirely new dimension to the analysis of gender equity. The economic argument is not the only argument in favour of the need to involve more women in innovation, it is also the vibrancy of the innovation process which benefits from having different views and several different approaches, often coming from minority groups, be these by gender or ethnicity.

One important aspect for understanding contemporary inventions and innovations is that they inherently relate to existing technologies and services, and also to existing R&D ecosystems and networks, as well as changes in them. Inventions are gradual, incremental and developmental, where the 'first time invention' is adjusted to the existing body of knowledge, or to existing devices or processes. In this process of adjustment, the actual innovation becomes embedded in the previous body of knowledge. Sometimes one piece – material or immaterial – of the new invention will make a big difference to the product, as has been often the case with smart phones and their development. Smart phones exemplify many of recent complexities in how the consumer interests and technological development go hand in hand.

Take Apple and the iPhone as one example here of the power of technologies and to show how technologies come about. Apple did not invent GPS (the global positioning system), but it made GPS an innovation through its innovative positioning as a key element shaping the future of mobile phones: Apple innovatively added GPS to their iPhone product and integrated GPS as a key part of the phone and its functions. Many applications and programmes use GPS or are based on its availability. This meant that the capabilities of the phone became exponentially greater because of the availability of and dependence on GPS. Combining GPS with the iPhone was an innovation which was accepted and applauded by the market and consumers alike (Worstall 2014). This type of scaling – exponentially creating new possibilities in existing fields through innovation – is typical for the new platform economy as will be discussed in Chapter 3 in more detail.

Much of the time we do not notice, nor are we able to grasp an invention or the processes through which it becomes an innovation and becomes an available product. Many high tech innovations go through a gradual type of invention process: combinations of patents, existing products or processes and new ideas lead to the innovation. Innovations are gradual upgrades of inventions, and are often invisible, ignored or even 'irrelevant' for consumers, even though the innovations could be foundations for other innovations. One, unexplored question in the innovation process is, where in organizational settings are inventions and innovations born? Furthermore, importantly, how do they become gendered? It is crucial then, that the context and organizational setting where inventions become innovations, and subsequently become parts of services and products, are taken into account when addressing gender in inventions.

The elements of the innovation process, such as performance, networks and enabling strategies differ by gender. The elements of the innovation process also differ by the age and level of education of the participants in innovation, both of which are intersected by gender in many ways (Whittington 2011; Ding et al. 2006; Naldi et al. 2005). A broad literature base exists which examines contemporary aspects of scientists and researchers involved in innovation development processes, as well as the related product value creation, marketing and branding of innovations. However, concerning innovators and inventors, that is, the actual persons involved, only a small amount of research so far exists that specifically looks at gendered questions in relation to the commercialization of science.

The commercialization of science has increased the need for boundary spanning between academic and commercial bodies of knowledge (Murray and Graham 2007). The reports and studies that elaborate the intersecting factors, such as age, gender and education, in addition to organizational settings, will be discussed in this and the following chapters. However, as gender is not a biological marker of the inventor, but a socially constructed reality and subject to social and cultural interpretations, it affects all innovations alike. For this reason, we will broaden our discussion from just focussing on gender differences and similarities in innovations to include the gendering of innovations in the following chapters.

There are several layers of organising involved in the way innovations become adopted and taken into use. One of the crucial layers are, of course, the inventors, who play a crucial role in defining and shaping their innovation. Some researchers (e.g. Jung and Ejermo 2014; Jones 2010) argue that contemporary inventions take relatively more time than in the past, due to the "burden of knowledge". The burden of knowledge refers to the idea that knowledge in general has become much harder to absorb over time, and for that reason the time it takes to invent has become much longer than in earlier periods. Jones (2010) argues that, in contrast to earlier historical periods, inventing in contemporary times requires a much longer incubation period, during which inventors learn and absorb an "ever-accumulating body of knowledge" (also Jung and Ejermo 2014: 111). Part of the 'incubation time' of the invention includes processing, adaptation and relating the invention to earlier data, material products and immaterial services. The complexity of knowledge grows over time and increases the need to coordinate new inventions with an ever growing, existing body of knowledge.

On the other hand, however, time may not be the most important element. Other factors such as user-experience and user-friendliness, the ability to appeal to the consumer in the case of consumer innovations, and ability to adjust to or disrupt an existing product and service may be a more crucial factor in innovations. In their analysis of Swedish inventors, Jung and Ejermo (2014) were able to take into account the age, gender and education level of the inventor, while also focusing on temporal trends between inventions, patenting and innovation. They used this data to test whether the 'burden of knowledge' theory holds true for the Swedish 'innovation climate'. Jung and Ejermo analysed a longitudinal database of patents filed between 1978 and 2009 in Sweden, with close to 60,000 matched patent-inventor-combinations. The numbers of patents and the number of inventors had both risen over time, as had the levels of knowledge involved and skills levels of the inventors (requiring more education than earlier).

Interestingly, the macro-level analysis over time of Swedish patenting data by Jung and Ejermo showed that the 'burden of knowledge' had actually lessened in Sweden since the 1990s. The reasons for this may not only be due to intensified R&D but also because of the motivation to innovate. This was backed up by the realization that the lead time for inventing among the highly educated had actually become shorter (Jung and Ejermo 2014: 115). According to their analysis, the overall propensity to become an inventor for those who have a Ph.D. in STAM (science, technology, agriculture and medicine) in Sweden is c. 2%, the figure being slightly higher for men and slightly lower for women. Furthermore, the share of those who participate in invention processes through their work in R&D in universities and in R&D intensive companies is also somewhat higher.

Several studies show that in academia women aim less for commercialization activities in technological and STEM fields in general (Ding et al. 2006, 2010). Reasons for this are many and they are explored further in the book. We will also ask whether the new digital technologies, new promises and possibilities of the big data and the platform economies create new possibilities for innovations and new types of product and service innovations. It is equally important to note that not all innovations are technology-based licencing and patenting-oriented inventions that need tech transfer. Increasingly social innovations, service innovations and creative work are part of the complex tapestry of economic and social progress of societies based on innovations. In that current development gender has new positions to take.

This all leads to the interpretation, by those who research invention patenting, that the number of inventors can be increased through changes in education policy, and through individual choices in the field of education. However, we ask whether the case really is such a simple equation after all. Is it only about the numbers? Can we reliably assume that when science and engineering education is more equally divided between the genders we will encourage more women into fields of study that tend to lead to innovation? Why then has this not taken place in countries, such as the Nordic countries, that already have a high level of equality, and where in the science fields, such as medicine, women often account for more than 50% of students and graduates? When we move from merely counting males and females to examining gendered dimensions of innovation processes and the contextualization of scientific knowledge, as in this book, a multidimensional and intersectional picture of innovation and gender begins to emerge.

In that picture, creativity in relation to gender, creative work and innovations takes new shapes. In the following chapters, we will discuss innovations, creative work, R&D and gender through literature and case studies with the aim to fill in unmapped areas on the canvass of innovation research with regard to gender.

REFERENCES

Abreu, M., & Grinevitch, V. (2013) The nature of academic entrepreneurship in the UK: widening the focus on entrepreneurial activities. *Research Policy*, 42(2): 408–422.

Abreu, M., & Grinevitch, V. (2017) Gender patterns in academic entrepreneurship. *The Journal of Technology Transfer*, 1–46, doi: 10.1007/s10961-016-9543-y.

Bath, C. (2014) Searching for methodology: feminist technology design in computer science. In E. Waltraud & I. Horwath (eds.) *Gender in Science and Technology: Interdisciplinary Approaches.* Bielefeld: transcript Verlag. 57–78.

Best, K., Sinell, A., Heidingsfelder, M. L., & Schraudner, M. (2016) The gender dimension in knowledge and technology transfer – the German case. *European Journal of Innovation Management*, 19(1): 2–25.

Bunker-Whittington, K., & Smith-Doerr, L. (2005) Gender and commercial science: women's patenting in the life sciences. *Journal of Technology Transfer*, 30: 355–370.

Bunker-Whittington, K., & Smith-Doerr, L. (2008) Women inventors in context: disparities in Patenting across Academia and Industry. *Gender and Society*, 22: 194.

Ding, W. W., Murray, F., & Stuart, T. E. (2006) Gender differences in patenting in the academic life sciences. *Science*, 313: 665–667.

Ding, W. W., Murray, F., & Stuart, T. E. (2010) From Bench to Board: Gender Differences in University Scientists' Participation in Commercial Science, Working Paper 11–014. Harvard Business School. Unpublished working paper. http://www.hbs.edu/faculty/Publication%20Files/11-014. pdf. Retrieved 10.11.2016.

Etzkowitz, H., Kemelgor, C., & Uzzi, B. (2000) *Athena Unbound: The Advancement of Women in Science and Technology.* Cambridge: Cambridge University Press.

Hellström, T., & Jacobs, M. (2005) Taming unruly science and saving national competitiveness: discourses on science by Sweden's strategic research bodies. *Science, Technology & Human Values*, 30(4): 443–467.

Hunt, J., Garant, J.-P., Herman, H., & Munroe, D. J. (2013) Why are women underrepresented amongst patentees?. *Research Policy*, 42(3013): 831–842.

Jones, B. F. (2010) Age and great invention. *The Review of Economics and Statistics*, 92(1): 1–14.

Jung, T., & Ejermo, O. (2014) Demographic patterns and trends in patenting: gender, age and education of inventors. *Technological Forecasting & Social Change*, 86: 110–124.

Link, A. N., Siegel, D. S., & Bozeman, B. (2007) An empirical analysis of the propensity of academics to engage in informal university technology transfer. *Industrial and Corporate Change*, 16(4): 641–655.

Lins, E., & Lutz, E. (2016) Bridging the gender funding gap: do female entrepreneurs have equal access to venture capital?. *International Journal of Entrepreneurship and Small Business*, 27(2/3): 347–365.

McKelvey, M., & Holmén, M. (2009) *Learning to Compete in European Universities: From Social Institution to Knowledge Business*. Cheltenham: Edward Elgar.

Naldi, F., Luzi, D., Valente, A., & Vannini-Parenti, I. (2005) Scientific and technological performance by gender. In H. Moed, W. Glänzel, & U. Schmoch (eds.) *Handbook of Quantitative Science and Technology Research*. Netherlands: Springer. 299–314.

Murray, F., & Graham, L. (2007) Buying science and selling science: gender differences in the market for commercial science. *Industrial and Corporate Change*, 16(4): 657–689.

Poutanen, S., & Kovalainen, A. (2013) Gendering innovation process in an industrial plant – revisiting tokenism, gender and innovation. *International Journal of Gender and Entrepreneurship*, 5(3): 257–274.

Poutanen, S., & Kovalainen, A. (2016) Innovating is not of the spirit world – depicting a female inventor's unique path with materiality-friendly gender concepts. In A. Alsos, U. Hytti, & E. Ljunggren (eds.) *Research Handbook on Gender and Innovation*. Cheltenham, UK, Northampton, USA: Edward Elgar.

Rosser, S. V. (2009) The gender gap in patenting. Is technology transfer a feminist issue?. *NWSA Journal*, 21(2): 65–84.

Rosser, S. V. (2012) *Breaking into the Lab: Engineering Progress for Women in Science*. New York: New York University Press.

Schiebinger, L. & Klinge, I. (2013) *Gendered Innovations. How Gender analysis Contributes to Research*. Directorate General for Research & Innovation. European Comission. Retrieved on 14th February 2016 from http://ec.europa.eu/research/science-society/genderedinnovations/index_en.cfm

Wacjman, J. (2004) *TechnoFeminism*. Cambridge: Polity Press.

Whittington, K. B. (2011) Mothers of Inventions? Gender, motherhood and new dimensions of productivity in the science profession. *Work and Occupations*, 28 (2011): 417–456.

Worstall, T. (2014) *Using Apple's iPhone To Explain The Difference Between Invention And Innovation*. Forbes, Apr 20, 2014. Retrieved on 14th february 2016 from http://www.forbes.com/sites/timworstall/2014/04/20/using-apples-iphone-to-explain-the-difference-between-invention-and-innovation/#8c9c40d56c46.

CHAPTER 2

Gender in Inventions and Innovations

Innovations do not take place in a vacuum, but are complex products of their time. Many innovations are based directly or indirectly on science and the everyday uses of science. Indeed, contemporary societies are permeated by science and the products of scientific endeavours. Innovation theory emphasizes the importance of vibrant clusters, be they regions or cities, and the scalability of ideas as primary reasons for a rich innovation culture. At the nexus of these clusters, we most often find universities, especially research universities that act as the fulcrum of innovation systems. How does gender become visible/invisible in the innovation culture? This and other similar questions will be discussed later in the book.

Science and research-based innovations are undoubtedly a major part of most innovation cultures. Even so, science is not one single entity, but rather it is a complex social phenomenon that takes different forms and shapes, all of which develop over time. Innovations may be tightly or loosely coupled with the ideas and inventions of science. Research into innovation ecosystems highlights the interconnected nature of individual components, with universities, sciences and scientists recognized as important cogs in innovation systems. However, using a machine metaphor assumes a kind of stability and a certain predictability in the machine parts functioning together and this is not necessarily in line with real life. We know from research into university technology transfer (e.g. Chen et al. 2016; Gray et al. 2014; Gianiodis 2014), that university technology transfer is a highly complex activity involving many stakeholders with

© The Author(s) 2017
S. Poutanen, A. Kovalainen, *Gender and Innovation in the New Economy*, DOI 10.1057/978-1-137-52702-8_2

competing interests and that scientists' participation in the process is crucial. Work-related aspects in science become critical for the success of the innovations, and it is in these work-related aspects that gender and gender-related effects become visible (e.g. Smith-Lawton et al. 2014; Mavriplis et al. 2010). Work-related aspects may relate to the contents of the actual work but also to the arrangements surrounding the work and range from formal salary systems and reward mechanisms to general working arrangements, decision-making processes, work-place cultures and networks which are made available through work-place culture.

This chapter takes up the issues of gender in inventions and innovations by establishing the historical roots of gendering in innovations, by discussing the recognized signs of inventions and innovations – namely the patenting, and by opening up the discussion on the key elements of innovation, science and gender.

2.1   GENDERING EARLY INVENTIONS

Even if we find it difficult to think about contemporary life without such revolutionary innovations as mobile phones or electric light, not all inventions evolve over time or are as enduring or as game-changing as Thomas Edison's long-lasting filament lamp and Alexander Graham Bell's telephone (Dummer 1997). A noticeable number of inventions also die off, never get patented or become non-productive, such as those shown in an analysis of patent failures by Bessen and Meurer, who analysed a wide range of empirical evidence focusing on history, law and economics. Their findings are stark: while patents do provide incentives for companies and societies alike to invest in research, development and commercialization, for most businesses today, patents also fail to provide predictable property rights. Instead, patents produce legal disputes, the costs of which can outweigh the positive incentives for patents (Bessen and Meurer 2009; Murray and Stern 2014).

Inventions may have been easier to recognize in the past, in contrast to nowadays in the contemporary world, but the path to inventions and to innovations has seldom been straightforward. Likewise, in the past, inventions also developed through shifts and changes just as they do today. Let's take an illustrative example from everyday life in the past that extends to current everyday life: the washing machine. The washing machine is a perfect example of the gradual attachment of new ideas and parts that came to be added to previous iterations and models. The laundry was women's exhausting work, most often with the heavy carrying of water, heating it and then washing, scrubbing, often with soap made of lye (Landau 2006). The needs-based invention of the washing machine and its development into an electrically powered machine coincided with several factors that were favourable for the invention and increased the demand for the product. These included the rise in the general standard of living, the development of cities and improvements in living conditions together with changes in the size of apartments, as well as changes in the structure of the labour market including the availability of servants and the shift from servant work to self-service.

In the case of early washing machines, the evolution of a new product occurred as the scrubbing board was replaced by the first cylinder machine. This development was helped along by the early market success of the invented prototype. The concept of using a rotating drum to clean dirty laundry was patented by an English inventor in 1782, and this was followed by a patent by James King in 1851 in the USA, whose machine also had a drum – which although was still hand powered and resembled a modern machine (Zmroczek 1992). The improvement in the efficiency of laundry work and its impact on the use of time devoted to household work were considerable (e.g. Zmroczek 1992; Bose 1979). Steady improvements in laundry devices took place from very early on and gradually changed the device. The different versions soon developed into dominant models in the early years of the twentieth century.

The intimate connections between the historical development of household appliances and gender are considerable and varied, and become manifested in the development of the washing machine: in 1874 William Blackstone, a merchant and manufacturer of corn planters, built a birthday present for his wife (Landau 2006). The machine was an early version of the modern washing machine. It was designed for convenient use in the home. The machine met growing demand, and the manufacturing of farming machines gave Blackstone the raw materials, tools and equipment needed for these new types of innovations.

The early Blackstone washing machine from 1880s consisted of a wooden tub with a handle, which moved a pad with small wooden pegs inside the tub (Landau 2006). By moving the handle, dirty clothes were scrubbed by wooden pegs in the water (Maxwell 2003). The machine functioned very well, and in five years' time Mr. Blackstone's wife's birthday present became a huge market success. The market was so large in fact that Mr. Blackstone began to build and sell his washing machines full time, abandoning the manufacture of farming machinery. With the increase in sales, he moved his company to New York. In urban centres such as New York, where population grew rapidly in early 1900s, the number of households was growing while the average size of the household was shrinking both leading to an increased demand for laundry machines. With the change in demographics, a rising standard of living and the growing availability of apartments with electricity and indoor plumbing, new requirements for household appliances, such as washing machines, emerged (Larkin 1989; Pursell 1995).

Household chores, such as the need to store food safely and the need to wash dirty laundry at home, shaped technological developments and designs leading to refrigerators and washing machines and, more importantly, also shaped early consumer product markets by creating the demand for products that saved household labour and women's work. The importance of innovation in laundry washing was visible in the sky-rocketing sales and number of companies and products in the relatively undeveloped consumer markets of the early twentieth century. This is reflected by the fact that in the early twentieth century there were more than 200 washing machine manufacturers in the USA (AHAM 2016), all making incremental inventions to develop their brand of laundry washing machines, but also varying the product itself to stand out alongside increasing consumer demand. The importance of the laundry washing machine and its development in reorganizing household work was visible in the sky rocketing sales and high number of companies and products in the relatively undeveloped consumer markets of the early twentieth century. The entry of the washing machine companies to the consumer markets varied, and it is interesting that most of the early innovation development for new home high-tech appliances of the early twentieth century, such as refrigerators and washing machines, took place in the USA and the UK.

The gendering of high-tech household products of their time, such as the washing machine, had already taken place very early on in the history of these household products and innovations. The Thor washing machine was one of the early successes, with the name of the machine referring to the masculine and powerful Viking god of thunder, lightning and electricity. The Thor washing machine was a drum-type washing machine with a galvanized tub and an electric motor, but with a water supply that did not necessarily require pipes (Esporta 2013). The Thor was one of the first inventions where the arrival of electric power underlined the superiority of powered machines over earlier manually operated versions. A patent was issued for the Thor machine in 1910 (Pursell 1995; Esporta 2013).

Initially washing machines were a side-product for many, more important farming machinery companies. The first washing machine with the name 'Pastime' was also a side-product of the farming machinery company Maytag, which produced farm machines for threshing, band-cutting and self-feeder attachments invented by one of the founders of the company in early twentieth century (Esporta 2013). Manufacturing farming machinery expanded into making household machines during the agricultural low season in the winter time. The outcome was machinery that used parts intended for farm machines, and had the relatively robust look of a modern industrial appliance. The robust outlook and differences in the product development were largely because some of the early washing machine producers were in farming device production with side brands in home appliances, such as washing machines, while others were exclusively developing and innovating new versions of home-based laundry appliances for growing markets.

Apart from refrigerators and washing machines, electricity also led to many other household devices, such as the electric stove and the electric iron, both of which employed innovations based on an innovation in engineering: resistance heating. Further innovations included devices such as the electric vacuum cleaner which was built around a small motor. Changes in household composition and family structure, as well as the loss of servants and a more active role for women in the household, in addition to the availability of electric devices as a labour saving solution to household problems, were all reflected in the General Electric advertisement from 1917, which talked directly to consumers: "Don't go to the Employment Bureau. Go to your Lighting Company or leading Electric Shop to solve your servant problem" (Greatest Engineering Achievements 2016).

In these early periods as society became industrialized, examples from everyday life show how in almost all cases, innovations were built according to two kinds of logic: the power of household logic – production mainly for own household needs – was overcome by market logic – production for the markets. Although the gendered division of household work within the family, as well as the invisible reproductive work of wives at home, gave impetus to early inventions such as the washing machine, rather early on market logic replaced household logic in the invention and innovation development work done in companies. The household logic of the early inventions created appliances as technical solutions and improvements to address household problems and needs. While this sometimes led to appliances and designs which were not as handy in use nor were pleasing to the eye, most were practical and fit-for-purpose.

With the growth of industrial production market logic replaced household logic in production and shaped technological development from very early on. Market logic directed consumers' desires and wishes towards newer models and ever-evolving technologies. With high demand and the market logic directing the development of the product, the washing machine rose in price, and became a more standardized household appliance for the maturing markets of the 1930s. The first automatic washing machine was introduced in 1937 by Bendix in the USA (Museums Victoria 2016; Esporta 2013). The underlying technology of the machine has not changed since, even though incremental inventions in electronics and digital control have increased the number of potential uses for the machine.

The early Maytag electric washing machine from 1911 exemplified two things: first, that inventions can take place through gradual and steady product improvements as in washing machines, and second, that gendered innovations moulded the machinery itself. These lead to an essential third point which concerns how paradoxical technological improvement increased the gendered division of work and household tasks. Representing power, 'Thor' – the ancient Viking god – created a powerful image of help and sustainability and was undoubtedly meant to substitute domestic service, which declined rapidly from 1900 to 1950 (Bose 1979). Societal transformation influenced the ways inventions became part of the household and its division of work. The attraction of technology took place hand in hand with the rise in the standard of living, the installation of electricity and piped water, increasing purchasing power and more varied consumption opportunities for households. Paradoxically, gender-specific innovations that 'freed' women from household work also led women towards consumerism and service-oriented innovations and tied women to the home anew. Research has shown that the changes in women's division of time between home and work cannot be explained by technology itself. Rather, as shown by Bose (1979), technology has also increased – and not diminished – women's role in the household as a second order effect of technological innovation.

The washing machine exemplifies both a 'game-changing' invention that has greatly transformed the use of time in mundane everyday life and also affected the gendered household division of work and time. At the same time, a game-changing invention (as the washing machine was in its own time) gives rise to a multitude of other inventions and shapes the ways work becomes further gendered. With industrialization, inventions greatly increased the efficiency of machinery across the board; consequently Fordist production and piecework in factories increased and became the dominant mode of industrial production (e.g. Dunaway 2014). At home, the gendering of household work was assumed to change with the arrival of new machinery and with the decline in the number of employed household servants and temporary help. Despite the assumed entry of technology into the household and the assumed masculinity of the technology, the use of the machinery in the household did not change gendered household work and the division of household work by gender as such – women are still more likely to deal with the laundry than men are. Paradoxically, new household inventions made household work even more gender divided: the washing machine did indeed replace the ever laborious handwashing of laundry at home, but as the new machine was usually placed in the kitchen, or in the communal laundry room of the building, the space also under-lined the stability of the gendered division of labour in the household, despite the time-saving prowess of the new machines. Simultaneously, the new machinery stabilized the reproductive function of the household, and did not move this reproduction entirely into the markets.

2.1.1   Case: How the Draining Closet for Dishes was Invented

The heavy years of the Second World War required women's labour on all fronts in all countries that were involved. As in most countries during the war, also in Finland women replaced men in factories: during the Second World War the majority (51% in 1943) of the industrial work-force in Finland were in fact women because most men were on active duty. During the war years, women took care of the home, industrial work and farming all over Europe. The rapid change in the women's formal labour force participation and, more generally, in their industrial work participation, is visible in the statistics: before the war the share of women in the industrial work force was 8%, and during the war years 51% of all those working in industry (Statistics Finland 2016). Women con-stituted a war-time labour reserve, which in Finland, unlike in most European countries, did not fully return to home and to household work following the conclusion of hostilities. This aberration from the usual pattern was caused by a situation where rapid economic recovery coincided with increasing demand for labour (Kovalainen 1995).

One reason Finnish women were needed in the workforce after the war was due to the war reparations that the country was liable to pay following the peace treaty between Finland and the Soviet Union. Additionally, much of the reparations were to be paid in the form of products, machinery and so on. To serve this truly immense purpose, all efforts were sought to make household work – which formed a second working day for most women – more efficient. Considering this historical period, it is very interesting that women's invisible household work was made a priority target for formal activities and awareness campaigns, such as those run by the Work Efficiency Institute in Finland. One of the pioneers in this field at the Work Efficiency Institute was Ms. Maiju Gebhard (Pulma 1984).

Ms. Gebhard (1896–1986) was a teacher of home economics at the Orimattila Household Institute and later in her career, she became a department head at the Work Efficiency Institute in Helsinki, Finland. She came from a well-off family, where her father was a professor and her mother was active in politics and a member of parliament. Ms. Gebhard had studied home economics in Sweden and graduated as a teacher in 1919, whereupon she returned to Finland. During the years she had studied in Sweden, Finland had become independent, and the Second World War was some way off. Nation building was one of the national priorities and for that purpose the Work Efficiency Institute was established. Ms. Gebhard was driven forward in her work by a strong belief that the best help for the less well-off was through education. At the Work Efficiency Institute Ms. Gebhard led research, education and even public awareness campaigns. In that role she developed several improvements which rationalized work in the home as well as improved time use, particularly saving women's time from household chores such as time spent in kitchen (Gebhard 1947; Valtonen 2014).

When studying in Sweden Ms. Gebhard had seen a removable wooden frame on the kitchen table in the household where she lived, where clean, wet dishes were put, then dried with a dishtowel and further placed in the cupboards and cabinets. Gebhard, in her strive to rationalize invisible and unpaid household work, started to consider how such a chore might be rationalized. Gebhard's innovation was the dish drying cabinet and her invention was to install a wooden frame with wire or wooden racks into a cabinet equipped with doors. The washed, still wet dishes could be lifted out of the water directly onto the rack in the cupboard with the remaining water dripping into the sink (or into a tray under the rack); the dishes would dry by themselves behind the closed doors. After a while, the dishes, which had dried in the airy cupboard, could be used directly. This saved considerable time when one or two phases were removed. Industrial production of dish drying

Thanks to Maiju Gebhard Finnish kitchens have integrated dish drying cupboards above the sink. There are always delicate or large dishes that need handwashing and drying in an airy cupboard.

(Picture: Puustelli Group).

Used with permission

cabinets began after the war with the Finnish Enso-Gutzeit company producing racks with a plastic coating. Since then, kitchen manufacturers in Finland have included dish-drying cabinets with their products (e.g. Työtehoseura 2016).

2.1.2   Case: How the windshield wipers were invented

Imagine driving your car without properly functioning windshield wipers in heavy sleet on a cold winter evening. What would you do to clean the windows of your car? Most probably stop the car to wipe the windows clean or stick your head and hand out of the side window to do the same while driving slowly. This was exactly the case 120 years ago when no windshield wipers existed in the first cars invented. It actually took quite some time before the 'window cleaning problem' was solved. For many years, the most usual way of cleaning the windows was for the driver to stick his head out of the window and clean the car window by hand. An alternative way was to drive with the windows of the car open to stop the frost from dimming the windows entirely.

In the winter of 1902, Ms. Mary Anderson of Birmingham, Alabama, visited New York and on her visit she took a trolley car. In the carriage, she observed the driver with the panes of the double front window open due to the difficulty in keeping the windshield clear of falling sleet (Slater 2014). She was very bothered by this situation and after returning to Alabama from New York, Ms. Anderson, who was a well-off woman, hired a designer to draw a hand-operated device which would keep the automobile's windshields clear at the wintertime. Her intention was to create a windshield blade that would connect itself to the interior of the car, allowing the driver to operate the windshield wiper from inside the vehicle. Ms. Anderson was so sure of her prototype that she even hired a local company to produce a working model. And so the everyday gadget found today in all cars was invented in 1903.

Ms. Anderson applied for a patent for the invention. The patent was granted to her for the windshield wipers in 1903 for 17 years. For several years, Ms. Anderson tried to sell her patent to the car industry, only to hear that it was not needed. Finally, when the patent expired in 1920 (Slater 2014) her invention was available for use freely by the car industry.

2.2   GENDERED PATTERNS OF PATENTING

The standard account of the history of innovation and inventions is often presented as a story of rather straightforward scientific progress, knowledge accumulation and growth with an increasing volume of discoveries peppered with a few twists and turns to add interest. The central idea in the history of discovery and invention is progress; indeed, the conceptualization is strongly tied to societal progress. In this grand narrative, there is little space for women or for gender more generally. On the contrary, science and technology have been considered a means of domination and not of liberation by many feminists (e.g. Hasse 2008). Indeed, for a long time formal science excluded women from its institutions and discriminated against those few who were admitted (e.g. Trask 2014).

Moving from historical development of inventions to the contemporary context, it is a well-established fact that pervasive and well-known gender inequalities in science, research and technology prevent economies from functioning optimally (Ghiasi et al. 2015). Even though the gender imbalance in terms of numbers of positions in research is well documented, there is a need to investigate how changing structural and cultural conditions in the research system influence the career progression of women and men. Previous studies have illustrated that innovation is gendered in at least three dimensions: (a) the gender labelling of the economic sector, (b) the definition of the innovation concept and (c) the innovator (e.g. Nählinder et al. 2012). In the following chapters, we will investigate some examples of these three dimensions. Many times these three aspects are also inseparable from each other.

Innovation in manufacturing is typically envisaged as being research-based activity. In particular, the increased emphasis on "excellence" and new demands for competitiveness, innovation and accountability require a gendered analysis of its effects. The innovation field, including private and public innovation, shows similar gender patterns (e.g. Nählinder et al. 2012; Nählinder and Tillmar 2013). These gendered patterns become more transparent in the research and innovation landscape, where for instance patenting is used as a measure, and is an established proxy for innovation activity. In the following, this chapter sets out to explore the current innovation landscape, and the most established phenomenon of invention and innovation activities: patenting.

The gendered disparities in science fields and careers are well known when it comes to the so-called leaky pipeline (Schiebinger and Klinge 2013; Blickenstaff 2005), career ladders (Sandberg 2013; Lindberg et al. 2012), and the persistent gender inequalities in hiring, funding and in scientific production (e.g. Shen 2013; Meng and Shapira 2011). However, the present knowledge of gender disparities in science is often monodisciplinary and limited by national settings and the range of activities measured. The disciplinary limitations are understandable, as for example, recruitment practices and publishing patterns vary greatly. A recent global comparison of women's and men's publications across countries and disciplinary fields with close to 5.5 million published papers in the Thomson Reuters database reveals that women publish with fewer citations, lower journal ranking levels and with fewer first authorships (Lariviere et al. 2013; also Bentley 2011). How might one explain this persistent difference?

Using existing data on women innovators in terms of patents filed and registered is one way to analyse the extent of activities and how women relate to and are part of inventions and innovations. Patents are considered to be markers of success and are highly important since through them one can claim ownership to a specific invention, and likewise, patenting functions as a seed for innovations (e.g. Stuart and Ding 2006). Patenting marks establish ownership which can have value years after the patent is granted. Most often, several steps are required for an invention to become a patented invention and processed onwards into an innovation (e.g. Yusof et al. 2014; Wihlman et al. 2014).

Patent litigation is one part of the contemporary innovation system, even though the likelihood of a patent becoming involved in litigation during some point of its life cycle is only 1–3% (Harhoff 2011). Continuously, a large number of patents are issued to protect technologies vital to devices such as smartphones, for example as a part of the so-called "smartphone patent wars" where leading manufacturers are trading legal suits over patent infringements, with some involving features such as double-tap or pinch to zoom that are everyday features in phones. Innovation is not only about successes and patenting but also about failures, hindrances and competition over intellectual property rights (e.g. Lungeanu and Norshir 2015).

Patents are also part of the financing system, and patent systems in Europe differ from the US system in several crucial respects. Two of the most marked differences are with regards to court proceedings, which in Europe are far less costly than in the USA, and in the quality of patent examination, which in Europe is higher than in the USA, for example.

Inventions of the twenty-first century are outcomes of processual and deliberate developments and changes. The inventor-entrepreneur who orchestrates and manoeuvres the idea from invention to innovation needs to work as meticulously and have at least as many personnel as Edison in his Menlo Park laboratory did in 1878 (Hughes 1999), which was called by Edison his invention factory with its own boarding house for single workers and underground system (Sproule 2000). Most inventions and patenting today are thus about gradual and processual improvements to the existing body of knowledge. The intellectual history of innovations is a history of variety. It is crucial to ask the question, as it is asked in this book, why did innovation come to be defined mainly as technological innovation (Godin 2008, 2015; also Berkun 2010).

As innovations are also generally understood as commercialized innovations only, the intellectual property rights are at the focus. However, there are gaps in the knowledge. For example, intellectual property rights are not often main point of focus when entrepreneurship is analysed (e.g. Autio et al. 2013). The reason for this is that in entrepreneurship research the focus is in the puzzle of new innovations and their entrepreneurial use, and not the IPRs. However, the IPRs are at the core of the scientific inventions and innovations today (Mui and Carroll 2013). Patenting holds the key to most, new, technological and science-based scalable innovations, and identifies the agencies behind the inventions. In the following, research focusing on patenting and gender is summarized.

On a global level, the gender of inventors has been studied to some extent, most often with large data sets and with standard socio-economic variables such as age, sex and education. In the USA the gender gap is, and has long been, rather rigid and wide in patenting: some estimates give 8% as the share of the patenting done by women (Hunt et al. 2013; Abreu and Grinevitch 2017). Frietsch et al. (2009) conducted a study on Europe-wide patenting patterns by men and women. In their study, the share of patents filed by women ranged from 2.9% in Austria to 14.2% in Spain. In a large analysis of patents by female researchers at the EU level by Busolt and Kugele (2009), the low percentage of female researchers within EU member states was accompanied by an even lower percentage of female inventors (Weber et al. 2014).

A consistent finding throughout the literature is that most women as inventors are to be found in the pharmaceutical industry and technology sector, followed by chemicals, and with the smallest number of women inventors in the machinery and mechanical engineering fields (Jung and Ejermo 2014; Naldi et al. 2005; Kugele 2010). This result resonates with patterns in the segregation of university education by gender: almost universally, and especially in western societies, fewer women choose engineering as their major in comparison to medicine. The reasons for this are complex, ranging from work-related ideas, assumptions and stereotypes, to actual discriminatory cultures. These will be discussed in the next chapters.

The gender gap in inventions and innovation appears to hold internationally: irrespective of the country, women obtain significantly fewer patents than their male counterparts (Rosser 2009; Simard and Gammal 2012). The lack of women in SET (science, engineering and technology) and STEM (science, technology, engineering and mathematics) fields in senior positions not only constitutes a gender gap in inventions but also a significant loss of potential in terms of new ideas, inventions and innovation. Especially computer science, mathematics, information systems and engineering all represent fields that have grown fast and show high demand for labour. Therefore, some research reports recommend so-called 'retention strategy' for organizations (Joshi 2014; Simard and Gammal 2012). A well-researched fact is that STEM education, research work and career pipelines are not straightforward upward-leading mobility ladders or one-directional tubes, but rather winding routes with side steps and even metaphorical leaks; this is irrespective of whether the research or business is the object of analysis (Polkowska 2013; Rosser 2012; Truss et al. 2012).

The metaphor of a leaky pipeline is used to describe gendered academic career development and the diminishing number of women on the way to the top. The metaphor also refers to the natural sciences' idea of a faulty pipeline, in contrast to an ideal, non-leaky, smooth pipeline. With current global trends in higher education, we may ask whether educational and career steps create a direct or straightforward path, or whether non-linearity has come to epitomize academic research careers. Research on women's careers in the STEM fields has emphasized the various stages in career development (e.g. Abels 2012). Recent research into careers in higher education report declining prospects for both genders concerning their career development in academic R&D (Schiebinger and Klinge 2013; Bozeman and Gaughan 2011, 2007) and provide a clear rationale to more closely investigate the ways organizations recruit their R&D personnel.

The reasons for the leaky pipeline in general are several, ranging from gendered research cultures, gendered promotion patterns and practices as well as gendered organizations, organization cultures and wider sets of institutions. The pipeline metaphor is based on the ideal of a smooth career flow and upward mobility from PhD to professorship. The use of the metaphor in academic career research functioned well when looking at career paths and the disparities of the 1980s, 1990s and early 2000s. Increasingly though, in the wake of market-based higher education activity with stronger competition and overflow in the education system, the metaphor of a pipeline does not accurately portray current neoliberal higher education institutions and R&D policies.

The analyses of existing leaky pipelines in academia list a number of reasons for the diminishing number of women in scientific careers. The reasons for leaving or opting out of academia range from individual factors to factors beyond individual control, such as processes and work place and organizational cultures, to list but a few (e.g. Schiebinger and Klinge 2013;

Schiebinger and Schraudner 2011). The leaky pipeline in general results in relatively small numbers of women entering those positions in science which emphasize basic research and publications over patents, or allow for long-term commercialization interests in academia. Studies analysing the reasons for differing positions show that the most common reason for the gender-patenting gap arises from the fact that women do not get to be in charge of research groups that actively work on inventions leading to patents (e.g. Campbell et al. 2013; Whittington 2011). One key aspects here is the discrepancy between the organizational ideal worker and the actual resources of women and men working in the organization.

It is known that gender, human capital, technical background, type of business and the social networks of the entrepreneur importantly shape decision making on invention activities and patenting, and in other, related types of work. The gender gap in high-tech entrepreneurship is particularly important to study as it constitutes a significant component in contemporary economic activity. Furthermore, the ways the stereotyped gender notions of femininity and masculinity shape and form organizational processes and practices in science and businesses can have an effect (Poutanen and Kovalainen 2013; Sang et al. 2014). In business, hegemonic masculinity is formed through gendered working practices and the agency of individuals working long hours in groups following the patterns of homosociality for example (e.g. Sang et al. 2014).

In science, severe competition creates gender discriminating practices, most often to the disadvantage of women in innovation activities. How much this then hinders the nation-state, for example, in using its full economic, cultural and other potential, is a concern for policy makers and numerous evaluation reports have addressed the low numbers of patenting women inventors. Regarding the future of women in science, although women are reaching top positions in science and innovation, gender equality is still far from being realized. Several trends in academia are currently changing the research cultures with increasing turbulence and competition (e.g. Hauge 2016; Deem et al. 2015). These trends transform the academia and also change the picture of research work and careers across generations.

2.3 SCIENCE CHANGING PATTERNS?

Patents are one of the lifelines for inventions in terms of scaling up towards innovations and commercial products and services. Although innovative activity is difficult, if not impossible to measure, one structured way to investigate it further is to analyse the patenting behaviour. Patents facilitate new technologies and products and when successful, they bring in wealth and they grant ownership rights to inventors.

Data on patenting activity by gender is limited in many countries as the data does not necessarily differentiate the gender of the inventor/s when the patent is registered. The matching of the patent data with existing data sources gives a proxy, and national surveys are common data sources used to examine patenting behaviour by gender (Ashcraft and Breitzman 2012; Perkmann and Walsh 2008). In the USA the analysis of the patent data by gender shows that between 1977 and 2002 the percentage and the number of patents in USA which included a woman inventor increased. In 2010 the number of patents with any women inventors listed was close to three times higher than the number granted in 1977 (Milli et al. 2016). The analysis of Swedish patenting data showed a substantial but gradually decreasing gender gap: the share of patents filed by women has slowly increased from 2.4% in 1985 to 9.1% in 2007 (Jung and Ejermo 2014). However, if we compare this change to shifts in other industrial sectors, we find that within patenting the narrowing of the gap between genders has occurred much more slowly than in other comparable economic activities, for example in the research careers of men and women.

In Sweden, where long time series data is available, the analysis of inventor-patent-data over time has shown, for example, that inventors have, in contrast to many other countries, become much younger over the years and better educated with more women entering this group, albeit slowly. The research results from the Swedish innovation ecosystem and patenting patterns put forward the question of the relevance of those policies that are now in place – not only in Sweden but elsewhere as well (Andersson et al. 2012; Nählinder 2013).

Any policy measure geared towards inventions and innovations should obviously take into account the vast variety of technologies in place and the related question of gender (Jung and Ejermo 2014; Hunt et al. 2013; Abels 2012). However, research into labour markets is also important. Those studies that have shown that the gender imbalance has been high in patenting (Frietsch et al. 2009; Sims et al. 2010; Meng 2016), in licensing (Klofsten and Jones-Evans 2000), and especially in academic spin-offs (Best et al. 2016), have also revealed the painstakingly slow progress of the gender balance in those fields. These imbalances are partly due to educational segregation. However, only a part of the skewness can be explained by women's under-representation in science and technology fields (e.g. Schmidt 2014; Rommes et al. 2012). Hence, the gender bias is even stronger within these areas than in research fields alone. Yet, knowledge about the underlying processes gendering this area and on the mechanisms and initiatives that may address such imbalances is scarce.

Even contemporary research on patenting and academic commercialization may still ignore gender, or understand gender merely as sex, an "individual characteristic", as stated by Perkmann et al. (2013). Growing attention is being given to the reasons explaining women's lack of engagement in patenting and research commercialization. Several papers state that male academics are significantly more likely to engage with industry than their female peers (Boardman 2008; Giuliani et al. 2010; Gaughan and Corley 2010). Overall, fewer academics are involved in commercialization in general. Lissoni et al.'s (2008) report of three European countries showed that in the total population of academics, approximately 4–5% of individuals had filed a patent. A roughly equivalent figure for the USA is provided by the Research Value Program, which states a figure of about 5% (Bozeman and Gaughan 2007). In the studies we reviewed the proportion of academics involved in patenting ranges from 5% to 40%, with the difference due to different sampling strategies, illustrating how patenting rates vary strongly between scientific disciplines, as well as research cultures and university cultures (Foster et al. 2015; Agrawal and Henderson 2002). Research cultures within disciplinary fields also vary between universities and countries.

The possible reasons for women's underrepresentation among patent holders have been analysed with the help of several national data sources, such as the national survey of college graduates (Hunt et al. 2013). The reasons for gendered disparity in patenting vary from individual reasons to institutionally embedded reasons (Hunt et al. 2013). Innovation is difficult, if not impossible to pin down or measure at the individual level, and so patents are used as a proxy. Even if not all innovations get patented, researchers assume that patenting correlates relatively highly with unpatented innovations (Whittington and Smith-Doerr 2005, 2008; Nager et al. 2016) and vary between disciplinary fields, being highest in technology and in biosciences (e.g. Ladd 2014; Lee and Motzkau 2012; Knorr-Cetina 1999)

Given the segregated education patterns for women and for men, it can be assumed that there is a difference, but it is clearly surprising that in USA, only 5.5% of commercialized patents are held by women, and women patent at only 8% of the male rate (in 2003). Career differences explain some part of these startling results. Sugimoto et al. (2015a; 2015b) differentiated patents by gender and by site (i.e., firm, individual and university), and the results showed the highest female patenting activities in universities (at 11%, compared to less than 8% in firms). Since 2003 in a period of ten years, the share has risen in USA among women in academia. In Europe, women's patenting figures vary: the highest shares are in Spain and France (12.3% and 10.2% respectively) and the lowest shares are in Austria and Germany (3.2% and 4.7% respectively) (also Ashcraft and Breitzman 2012; also, 2007).

The organizations involved matter for gendered disparities in innovation and patenting. The sites and places for patenting differ in their gender divisions, opportunities and possibilities. The sites for patenting are highly contingent on the field of innovation. Despite the 'leaky pipeline' discourse, it seems that when patenting is analysed, academia is a more productive and better place for women than business R&D departments. Recent research results show that, in every technological area, female patenting is proportionally more likely to occur in academic institutions than in corporate or government environments (Sugimoto et al. 2015a). The gendered disparities become visible when the technological impact of women's patenting is compared to that of the patents men had filed: women's patenting has lower technological impact than that of men (Sugimoto et al. 2015b).

When analysing women's so-called underperformance in patenting, Hunt et al. (2013) also found significant predictors for patenting – both for men and for women. Patenting in the USA is higher for men than for women, higher for researchers with more publications, and with more co-authors per publication, and for those with company scientists as co-authors. Continuing this line of enquiry, those researchers who are in universities with a strong culture of patenting also have higher patenting activity than peer universities and consequently rank higher in patenting statistics (Hunt et al. 2013). This illustrates that culture is one key factor in determining who places patents and where they do so – it is also clear that nurturing such a culture takes time.

More qualitatively oriented studies show that academic women fail to make early contacts with industry and for that reason fall behind men in developing the skills they need (e.g. Murray and Graham 2007). Academic women also have smaller, denser networks with fewer industrial contacts than men (Weber et al. 2014; Waltraud 2014; Hunt et al. 2013).

The gender gap in terms of scientific impact has been relatively well documented (e.g. Lariviére et al. 2013; Stephan and El-Ganainy 2007).

Given the low number of women involved in patenting, from the point of national economies it is crucial to increase the number. However, as the impact of patenting is another factor counting towards scientific impact, perhaps the most important contribution towards innovation is to see patents being cited too. References to patenting are growing in importance in general, but little is known about the technological impact gap, that is, how many times women's patents are cited in other patents compared to the patents of men in the same industry-specific or subject-specific field.

If we look at the fields and types of patents women file, it is interesting to note in which fields of science women publish their patents. It has been suggested that when female inventors are involved, patents tend to have higher diversity. This is confirmed in studies where the gender division of patents filed and the classification of the patents are analysed. In terms of the number of International Patent Classification (IPC) codes assigned, women have larger diversity in patenting, and they also have more co-patenting than men (Sugimoto et al. 2015a; Meng 2016; Meng and Shapira 2011). Sugimoto and her colleagues have analysed gender disparities in patenting by country, technological area and by the type of assignee, using a large dataset over a 40-year period of time. Female inventors registered in the International Patent Classification (IPC) were more often associated with numerous IPC codes, irrespective of the type of assignee, suggesting higher interdisciplinarity in patenting by women than is the case with men. Even when the researchers controlled for the concomitant increase in IPC codes and the share of women over time, thus taking the change over time for both factors into account, the difference remained significant. In addition, female inventors over time had more co-inventions on average than male inventors (Sugimoto et al. 2015a, 2015b). Empirical results, however, also show that the share of patenting by women remains at a lower level in comparison to women's numbers in science, technology, engineering, and mathematics fields and professions otherwise. In contrast, women have authorship in c. 30% of scientific papers, including co-authorships with men (European Commission 2013; Whittington and Smith-Doerr 2008).

The recent analysis of the literature exploring the reasons for women's underrepresentation in patenting at USA notes that women's underrepresentation in key patent-intensive STEM fields, such as engineering, explains more of the patenting gap than women's underrepresentation in STEM fields in general (Milli et al. 2016). This, combined with the possible obstacles in the workplace environments and work-life policies and generally fewer networks, does not predict a rapid change in the patenting statistics by gender.

Patenting in its broadest sense refers not only to the invention itself but also to the capabilities to register inventions, manage patenting processes and deal with sometimes complex stakeholder issues. Most often, these tasks are done by brokers: in universities this is done by university technology licensing offices and in companies patenting experts are part of the commercialization process. Most universities' technology licensing offices also proactively harvest potential patents.

If patenting is the foundation in innovation activities, as is often stated, securing ownership of the activity for the future, it is clear that corrective measures to increase women's patenting will need to be implemented. There are several potential measures, which range from ensuring that organizational cultures are favourable and supportive of patenting, to gender-balanced research groups and teams, and to individual career related patterns and norms that see patenting as part of a successful career. All these measures bring the question of education to the forefront in the actions that need to be taken in order to better women's participation in patenting.

2.3.1   Case: EU Prize for Women Innovators, Increasing Attention to Gender and Innovation

The European Commission established a prize for women innovators in 2011, in order to raise awareness of women in innovative industries and enterprises with entrepreneurial and value creation potential. In its sixth year running the EU Prize for Women Innovators, the European Commission wants to give public recognition to outstanding women entrepreneurs who brought their innovative ideas to the economy and knowledge market. The aim of the prize is "to inspire other women to follow in the footsteps of the winners" (EU women innovators 2016), and demonstrate the new and already existing ways for women innovators and their careers to progress. Most of the top candidates and finalists so far are from the biological and natural sciences, engineering and medicine, not from the social sciences or humanities.

In 2016, nine European women scientists and entrepreneurs who had brought their breakthrough ideas to the market competed for the EU Prize for Women Innovators 2016. The competition, which is open to all European women scientists and entrepreneurs and has strict rules, resulted in 64 submitted applications in 2016. The finalists were selected by a high-level jury consisting of independent experts from business, venture capital, entrepreneurship and academia. The companies created by the contestants showcased innovations in a wide range of fields and sectors, with life science and ICT being dominant. The 2017 competition announces additional prize for young female entrepreneurs, the Rising Innovators award, along with the prizes for outstanding achievement of three women entrepreneurs who have brought an innovation to market (European Commission 2016).

The inaugural EU Prize for Women Innovators winner in 2011 was Dr Gitte Neubauer. Dr. Neubauer co-founded the biotechnology company Cellzome, which is a leader in the development and advancement of proteomics technologies which help to develop better targeted drugs against inflammatory diseases and cancer. The proteomics technologies Dr. Neubauer developed with her colleagues at Cellzome can be used in drug discovery from screening to selectivity profiling of compounds in different cells and also in patient samples. The technologies that Cellzome developed differed from other traditional methods at the time used in early drug discovery by assessing drug interactions with target proteins. Cellzome's technologies more closely represented the structures and processes found in a biological system. This resemblance gave scientists the opportunity to observe how candidate drugs affect intended, desired and non-desired targets in a close-to-physiological environment and thus had the potential to pinpoint possible safety issues earlier in the process than by other means. In 2012, GlaxoSmithKline plc (GSK) announced the acquisition of Cellzome, after nearly four years of successful collaboration. During the collaboration, GSK and Cellzome had developed two early stage research collaborations in immune-inflammation therapy. The disclosed price for the company, started by Dr. Neubauer, the first winner of the EU competition, was £61 million (US $99 million) in cash (EU women innovators 2016).

Given the complexity of the invention processes and the differences in the work surrounding inventions in the private sector and in universities, there are no quick fixes for increasing the number of women in patenting, or indeed, in the fields where patenting is done, whether this is in academic research or in business. Structural issues can lead to gender bias in science, and for their part, they can also lead to more transparent and fairer discussions of background assumptions (Longino 2002; also, 1994). A number of studies, such as the SHE figures, which focus on Europe, and the FP6 programmes on gender and science, confirm the development of a skewed gender balance in science, and increasingly so in innovations. Women as an 'untapped resource' (e.g. Hasse 2008) has become a political slogan with little consequences in actual science policy. Data from the SHE Figures 2013, a major EU publication that presents Europe-wide data on women in science from tertiary education through to the job market, show that there are nowhere near enough women at the top levels of science and research (European Commission 2013).

Personal choices in research careers are central, according to many investigations, and they for their part indicate how the researchers' paths diverge during their careers. Even more important is the question of how women are able to reach the positions in their research careers where opportunities for patenting exist. Within disciplinary fields, there are disciplinary-specific differences that have an effect on how and in what ways opportunities for patenting become available. One of the remedies the research brings forward repeatedly is the share of women in STEM education, especially in electric and mechanical engineering, which is the most patent-intensive field (e.g. Hunt et al. 2013; Whittington and Smith-Doerr 2008). Recent studies show that attempts to integrate more women in engineering have been only partially successful because gender disparities are also rooted in cultural associations linked to engineering, technology and masculinity (e.g. Ghiasi et al. 2015; Peterson Mcintery 2010; Fara 2004). The lower number of doctoral degrees held by women plays a surprisingly minor role and the most important determinant which explains the gender gap is the underrepresentation of women in patent-intensive fields of study and in patent-intensive jobs. Furthermore, globally, local and regional steps to remedy the problem have been taken through initiatives for women and networks among women in technology and other networks for women.

The research results over a long time span by Sugimoto and her colleagues show that women find more collaborative ways to patent, do more interdisciplinary patenting and work in a larger number of disciplinary fields than men do, but do not have similar access to the positions that men have in science and technology fields. This leads to the situations where women patent at less 'advanced' levels of technologies in comparison to men. Globally, the majority of patenting, close to 90%, is done by men throughout the scientific community, and in company R&D centres and corporations. Women are the minority in every technological area where patents are registered globally (e.g. Sugimoto et al. 2015b). The gendered disparity in patenting extends to a disparity in scientific paper authorships and citations. Women in engineering at universities have a higher share of female authorship compared to women in engineering outside of universities in any industrial sectors. Studies show that in co-authorship and collaboration networks in engineering, women occupy less central positions compared to their male colleagues (Ghiasi et al. 2015).

Educational segregation is one of the explanations for gender disparity in patenting and innovations. Masculine cultural aspects of engineering workplaces are another major factor that result in gender disparity (Etzkowitz et al. 2000), tokenism (Poutanen and Kovalainen 2013; Kanter 2000, 2001), and the in/visibility paradox at work (Faulkner 2009), especially in male-dominated fields such as engineering (e.g. Bijedic et al. 2016; Faulkner 2014; Poutanen and Kovalainen 2013; Aubert 2014; Alsos et al. 2013). Often as viewed as token women in the workplace, women may be visible, but in fact they may be invisible as engineers or as members of invention teams. Changing these invisible processes of gendering in work place cultures requires recognition of skills and competencies as well as a gender aware leadership. What is also needed is a more deeply rooted analysis of the masculinities of the engineering and technologies involved.

That the image of future employment and work tasks has an important role in selection of future direction of studies is affirmed in several studies looking into motivations in study selections. A longitudinal comparative study of gender differences in education and career choices in the USA, Canada and Australia found that the girls tended to choose careers in the biological, social and environmental sciences and in medicine, over the mathematically based sciences because they perceived the latter to be less people-oriented and to have less value to society. The 'attainment/utility' ("importance") values were more important for girls in their career choices and science choices than for boys (Watt et al. 2012). Gender socialization practices, the degree and type of early choice and specialization have an important role in the early selection phase, emphasizing the importance of the educational processes.

Some studies emphasize family influence in education selection. The experiences of innovators who have started their own businesses confirm and can identify the 'burden' of family traditions, when choosing their educational field (Kariv 2013). Several papers also indicate that social class supporting or hindering girls in their educational field has effects world-wide, but is especially detrimental for girls in developing countries (e.g. European Commission 2013). Barriers and support mechanisms range from the private to public sphere and from the national to the supranational. Thus the question of education and its role inevitably widen to include societal and cultural fields where policies become forefront issues.

The suggestions above evidently set tasks for both education policy – higher education included – and for other capacity building policies, such as innovation policies of the state and HR policies in firms. There are several nation-wide and even global initiatives already focusing intensely on addressing the matter; for example, there are numerous university initiatives to support girls and young women (these initiatives can be found in most high-ranking universities, ranging from the United States to Europe and Asia). Additionally, instigation of such policies needs to address both awareness training and different types of mentoring programmes. Outreach to high schools is one such route recommended for attracting and increasing the number of potential STEM students. Here also professional associations, such as the association of feminist economists, have run training and awareness campaigns. Finally, efficient outreach is possible through companies' own awareness campaigns and training programmes and most of all, through fostering an invention and innovation-friendly atmosphere at the workplace and in the organization (Poutanen and Kovalainen 2013, 2016).

2.4   WIDENING THE FIELD OF INNOVATIONS

For years, the field of women and innovation has often been framed with the rather straightforward questions of: "Why so few? Why so slow? Why so low?" (Ranga and Etzkowitz 2010). Progress in the numbers of women in innovation, as described earlier in this chapter, is slow, when measured by the number of women's patents and co-patents, despite prominent policy measures in some countries that boost women's participation in technology education fields, and in patenting. It may well be that the relevant questions posed by Ranga and Etzkowitz, which in fact address the persistence of relatively small numbers of women scientists, researchers and innovators, receive only partial answers without the proper policy analysis.

The answers to the questions of, "Why so few?" and "Why so slow?" are most often covered by arguments regarding women's limited participation and scarce opportunity for involvement in innovation activities, because of their lower enrolment in technological education and lower numbers in science, research, and high-tech sectors, as outlined earlier in this chapter. When the analyses is extended to research, it is in mechanisms such as research funding and disparities in awarding the funding where effects have been found in relation to the persistent gender divide. Establishing the actual cause and effect from amongst the multitude of features that are specific to the gender divide in academia may be difficult (Ranga et al. 2012; Sugimoto et al. 2015b). Furthermore, there are two salient effects which have been identified: the classical 'Matthew effect', where greater recognition is given to established scientists whose contributions are more readily accepted by the public and wider audience (e.g. Merton 1968; Ranga et al. 2012), and its counterpart, the 'Mathilda effect' which works against, and unfortunately often to the detriment of women scientists as shown by Ranga and her colleagues (Ranga et al. 2012). The disparity question relates to the so-called productivity puzzle (Cole and Zuckerman 1984; McKelvey and Holmén 2009): the continuing difference in research productivity between academic men and women both in USA and in Europe. Gender effects in research funding, and gendered elements in funding processes and instruments, have also been evidenced in many articles and research reports (Wennerås and Wold 1997; European Commission 2013; Lissoni et al. 2008; Etzkowitz et al. 2000). These produce complex effects in innovation and go some way to explaining the scarcity of women in innovations and patenting.

One of the main elements in evaluations of the workability of innovations is the peer-review process whereby the best innovations can be identified, promoted and funded. One of the major criticisms brought to the peer-review system is that it is a male-dominated, often networked, collaboration where power disparities over resource allocation and decision making exist (Ranga et al. (2012). Wennerås and Wold (1997) studied gender patterns in peer-reviews and funding procedures in Sweden at the end of the 1990s. They found significant discrimination against women in the process of awarding fellowships by the Swedish Medical Research Council. Since the Swedish publication, several assessments of gender bias in funding mechanisms have been launched in the USA and in Europe during the 2000s. While peer reviews function as gate-keepers for innovation research and research funding, it is difficult to draw conclusions, due to differences in disciplinary fields, procedures and funding mechanisms. The question of innovation lies at the core of the issue; but it is important to look beyond the traditional understanding of innovation as a set of technological and patentable activities, widening the field of innovation to include social innovations and collaborative work in innovation development.

Nevertheless, it can be asked whether the question is built partially on wrong assumptions of accumulative transformation and linear development, when it is possible that these are no longer necessarily available. Given the increasing competition in science fields, changing labour markets, the rapidization of science and innovation, and the transformation of academia and R&D activities because of global competition, there are indications of another structure different from linear development. This involves a greater change in innovation patterns and practices which concern not only the numbers of people who do innovation work but also the organization of innovation activities. Digitalization changes not only the ways the data is understood in innovations but also, and more importantly, the ways innovations are developed.

The possible barriers to innovations, presented by Kariv (2013), range from the individually oriented "self-direction to the conventional, avoidance of the innovative" to the family oriented obstacle of "family and social expectations", and "work-family conflict", to societally and culturally regulated "social stereotypes and biases" and to structural, labour market inequality – types of barriers (Kariv 2013: 270). While none of these barriers are separate, clear-cut or stable, they create a void and question whether innovation supporting policies are women-friendly or whether they target the relevant issues with regard to women's opportunities for innovation. The lack of multiple examples of innovation activities due to the lesser participation of women brings forth the need to open up the "black box" of innovation and of innovating processes. The clear inflexibilities that exist both in knowledge-intensive innovation and in research life are not easily changeable, for example, in an academic career coinciding with child-bearing years and thus causing breaks in working life, and a culture of long working hours. The gendered "separation of labour" in science, which refers to the segregation of science fields, with women better represented in biological sciences and medicine, and men in the physical sciences and engineering (Ranga et al. 2012: 14), is not easily changeable either.

2.4.1 Case: Gender and General Science and Innovation Awards

The Millennium Technology Prize for innovations was established in 2004 by the Technology Academy Finland, to be awarded to great, internationally recognized innovators, for their disruptive innovations changing our lives for the better. The Finnish Technology Foundation award, The Millennium Technology Prize is worth 1 million euro and it is awarded every second year to ground-breaking technological innovations that 'enhance the quality of people's lives', and that stimulate further cutting edge research and development. The prize is awarded to innovations that have been applied in practice and are delivering extensive innovation-based changes currently and in the future. The winners of the Millennium technology prize for innovations include Linux inventor Linus Torvalds and Tim Berners-Lee, who invented the World Wide Web, an internet-based hypermedia initiative for global information sharing in 1980s. The LED light inventor and 2014's physics Nobel laureate Shuji Nakamura is among the recipients of the reward (Technology Academy Finland 2016).

The bi-annual announcement of the Millennium technology prize recipients has never emphasized the gender of the award winner prior 2016. In fact, the receivers of the reward have been 'genderless', as most often is the case for science awards given to men scientists and innovators. Typically in the case of the male recipients of the science awards, the gender of the recipient is not mentioned in the media outlets or in the news. This is in contrast to the way women get celebrated when they receive an award. Gender is at the forefront of the media for those few women who are among the recipients of the awards and prizes in technology and innovation fields (Technology Academy Finland 2016).

Frances Arnold received the Millennium technology prize in 2016 for the interdisciplinary research on 'directed evolution'. She was the first woman to win the prestigious Millennium technology prize in its 12-year history. Arnod is a biochemical engineer and her research on 'directed evolution' has helped sustainable development and clean technology become available and usable in several industries. Arnold's research uses biology by manipulating DNA to modify the types of proteins known as enzymes, to make the chemicals to work in sustainable ways in products we use in our daily lives. Arnold's innovation has been used for example to manufacture pharmaceuticals such as treatment for type 2 diabetes (Technology Academy Finland 2016).

The odds for a woman – one out of six – to receive innovation-focused Millennium technology prize are higher than for a woman to receive the Nobel prize – approximately one out of twenty – even if these two prizes are not comparable. The Millennium technology prize emphasizes the requirement of practical implications of the innovation and directs the award away from Nobel profile of basic research orientation. Nobel Prize is awarded "to those who, during the preceding year, shall have conferred the greatest benefit to mankind", as stated in the original will by Alfred Nobel. Since 1901, only 5.6% of Nobel Prizes have been awarded to women (Nobel Prize 2016).

The Breakthrough Prize targets fundamental physics, life sciences and mathematics. Laureates for the Breakthrough Prize receive $3 million each in prize money, making each of the Breakthrough Prizes the largest scientific awards in the world. The Prize was established in 2013 by foundations established by influential couples behind Google (Sergey Brin and Anne Wojcicki), Apple (Mark Zuckerberg and Priscilla Chan), Alibaba (Jack Ma and Cathy Zang) and investor Yuri Milner and Julia Milner. Of the 2016 five laureates in life sciences, one recipient was woman. The physics award followed the modern idea of groups working for joint target and was awarded to five teams consisting of over 1.500 people (Breakthrough Prize 2016).

The Economist's Innovation award has followed a different logic by focusing on rewarding the proven innovations that have shown their value with the great impact on business and society. The idea to support proven concepts follows the business logic of success but also the idea of renewal of businesses through innovations. The Innovation award is given in eight fields: bioscience, computing and telecommunications, energy and the environment, social and economic innovation, process and service innovation, consumer products, a flexible "no boundaries" category, and the corporate use of innovation (The Economist 2014). For the Economist, gender is not specifically brought up in similar manner as in other prizes presented above.

Women's hierarchically lower position, in comparison to men in academia and in business, predicts slow changes in the opportunities for participation in innovation activities and patenting. Unfortunately, the gender dimension of innovation is often considered to be subject to other elements in the process of innovating.

Science prizes, such as described above, shed light on other possible elements in the process of innovating. There are other prizes that emphasize gender. Such prizes recognize innovations but also careers in relation to innovation. Even if prizes separate and isolate individuals from research teams and networks, the visibility of individuals through recognition is important by itself. However, there exists a controversy regarding the focus on women in science among women scientists themselves. This is clearly articulated in an interview of molecular biologist, Dr. Genevieve Almouzni, who was the laureate of the 2013 FEBS/EMBO Women in Science Award. She is cautious about emphasizing the idea of "women in science" that many of the current science prizes convey. "I do not want my work to be recognized just because I am a woman. I want to be judged on the quality of my work and on that intrinsic quality only" (Horizon 2013: 3).

The comment above by the laureate of the prize reflects a rather typical view among scientists – women and men – who would like to think science is neutral territory and detached from gender and any gendered effects. According to this view, gender has no presence in science or in its reward system, and thus gender should not receive any extra attention in the practice of science or in the making of innovations. This type of thinking decouples gender and science. It builds on the assumed neutrality of science, and most importantly, positions gender as inferior to science and an outlier in relation to the science in the making. Most often, this type of thinking reflects the processes in token positions where visibility and standing out due to tokenism can become perceived as highly problematic (Poutanen and Kovalainen 2013).

This chapter has highlighted some issues that have a role when explaining the current lower numbers of women in the field of innovations and the slow pace of the change that is anticipated to speed up in the future. The deep divisions in research and scientific work in relation to innovations require also the widening of the rather narrow concept of innovations adopted. In the following, the aim is to address the question from the point of view of the new economy and women's innovative activities in the new economy. Often these barriers are described in a rather universal manner and are related to the 'old economy' with hierarchical structures, rigid leadership and steady positions and it is often assumed that these elements will disappear with the transformation of the economy.

The increasing attention paid to the persistent gender disparity in the innovation landscape has brought about policy concerns and actions intended to increase the number and share of women both in knowledge and technology transfer and in those science fields where invention and patenting are most common in academia (Schiebinger and Klinge 2013; Best et al. 2016). Understanding the gender consequences of this is challenging because innovation studies do not usually focus on individuals, but on innovations and inventions, companies included (Poutanen and Kovalainen 2013). Very often, the structural barriers, work place level and team level analyses of innovation processes push gender aside. Do the aforementioned barriers exist in a similar manner in the new economy, or are they detached from the new production modes? Such questions are the focus of the next chapter.

References

Abels, G. (2012) Research by, for and about women: gendering science and research policy. In G. Abels & J. M. Mushaben (eds.) *Gendering the European Union: New Approaches to Old Democratic Deficits*. Gender and Politics Series. London: Palgrave MacMillan. 187–207.

Abreu, M., & Grinevitch, V. (2017) Gender patterns in academic entrepreneurship. *The Journal of Technology Transfer*, 1–46, doi: 10.1007/s10961-016-9543-y.

Agrawal, A., & Henderson, R. (2002) Putting patents in context: exploring knowledge transfer from MIT. *Management Science*, 48(1): 44–60.

AHAM (2016) Association for home appliance manufacturers. www.aham.org. Retrieved 4.10.2016.

Alsos, G., Ljunggren, E., & Hytti, U. (2013) Gender and innovation. *International Journal of Gender and Entrepreneurship*, 5(3): 236–256.

Andersson, S., Berglund, K., Gunnarsson, E., & Sundin, E. (2012) Introduction. In S. Andersson, K. Berglund, E. Gunnarsson, & E. Sundin (eds.) *Promoting Innovation. Policies, practices and procedures*. Vinnova Report VR 2012:08. Stockholm: Vinnova.

Ashcraft, C., & Breitzman, A. (2007) *Who Invents IT? An Analysis of Women's Participation in Information Technology Patenting*. Boulder, CO: National Center for Women & Information Technology.

Ashcraft, C., & Breitzman, A. (2012) *Who Invents IT? An Analysis of Women's Participation in Information Technology Patenting, 2012 Update*. Boulder, CO: National Center for Women & Information Technology.

Aubert, J. (2014) *Women Entrepreneur Revolution: Ready! Set! Launch!.* Bloomington, IN: Balboa Press.

Autio, E., Broström, A., D'Este, P., Fini, R., Geuna, A., Grimaldi, R., Hughes, A., Krabel, S., Kitson, M., Llerena, P., Lissoni, F., Salter, A., & Sobrero, M. (2013) Academic engagement and commercialisation: a review of the literature on university–industry relations. *Research Policy*, 42(2): 423–442.

Bentley, P. (2011) Gender differences and factors affecting publication productivity among Australian university academics. *Journal of Sociology*, 48(11): 85–103.

Berkun, S. (2010) *The Myths of Innovation*. New York: O´Reilly Media.

Bessen, J., & Meurer, M. J. (2009) *Patent Failure: How Judges, Bureaucrats, and Lawyers Put Innovators at Risk*. Princeton: Princeton University Press.

Best, K., Sinell, A., Heidingsfelder, M. L., & Schraudner, M. (2016) The gender dimension in knowledge and technology transfer – the German case. *European Journal of Innovation Management*, 19(1): 2–25.

Bijedic, T., Brink, S., Ettl, K., Kriwoluzky, S., & Welter, F. (2016) Women's innovation in Germany – empirical facts and conceptual explanations. In A. Alsos, U. Hytti, & E. Ljunggren (eds.) *Research Handbook on Gender and Innovation*. Cheltenham, UK, Northampton, USA: Edward Elgar.

Blickenstaff, J. C. (2005) Women and science careers: leaky pipeline or gender filter?. *Gender and Education*, 17(4): 369–386.

Boardman, P. C. (2008) Beyond the stars: the impact of affiliation with university biotechnology centers on the industrial involvement of university scientists. *Technovation*, 28(2008): 291–297.

Bose, C. (1979) Technology and changes in the division of labor in the American home. *Women's Studies International Quarterly*, 2: 295–304.

Bozeman, M., & Gaughan, M. (2007) Impacts of grants and contracts on academic researchers' interactions with industry. *Research Policy*, 36(85): 694–707.

Bozeman, M., & Gaughan, M. (2011) Job satisfaction among university faculty: individual, work, and institutional determinants. *The Journal of Higher Education*, 82(2): 154–186.

Breakthrough Prize (2016) General introduction. www.breakthroughhprize.org. Retrieved 25.5.2016.

Busolt, U., & Kugele, K. (2009) The gender innovation and productivity gap in Europe. *International Journal of Innovation and Sustainable Development*, 4(2/3): 109–122.

Campbell, L. G., Mehtani, S., Dozier, M. E., & Rinehart, J. (2013) Gender-heterogeneous working groups produce higher quality science. *PLOS One*, 8(10): e79147. doi: 10.1371/journal.pone.0079147.

Chen, A., Patton, D., & Kenney, M. (2016) University technology transfer in China: a literature review and taxonomy. *The Journal of Technology Transfer*, 41: 891–929.

Cole, J. R., & Zuckerman, H. (1984) The productivity puzzle: persistence and change in patterns of publication of men and women scientists. *Advances in Motivation and Achievement*, 2: 217–258.

Deem, R., Kovalainen, A., & Poutanen, S. (2015) Words and Money – Ethnography of Science Evaluation in Austere Times. Paper presented at 4S Conference San Diego, US. November 11-14.

Dummer, G. W. A. (1997) *Electronic Inventions and Discoveries: Electronics from Its Earliest Beginnings to the Present Day*. 4th ed. Bristol, UK: Arrowsmith Ltd.

Dunaway, W. A. (2014) Bringing commodity chain analysis back to its world-systems roots: rediscovering women's work and households. *Journal of World-Systems Research*, 20(1): 64–81.

Esporta (2013) Evolution of washing. www.esporta.ca. Retrieved 21.6.2016.

Etzkowitz, H., Kemelgor, C., & Uzzi, B. (2000) *Athena Unbound: The Advancement of Women in Science and Technology*. Cambridge: Cambridge University Press.

European Commission (2013) *She Figures 2012: Gender in Research and Innovation*. Brussels: European Commission. Available at: http://ec.europa.eu/research/science-society/document_library/pdf_06/she-figures-2012_en.pdf.

European Commission (2016) EU Prize for Women Innovators. http://ec. europa.eu/research/innovation-union/index_en.cfm?section=women-innova tors. Retrieved 10.12.2016.

EU women innovators (2016) Accessed 3.4.2016 at http://ec.europa.eu/research/innovation-union/index_en.cfm?section=women-innovators&pg=who.

Fara, P. (2004) *Pandora's Breeches: Women, Science and Power in the Enlightenment*. London: Pimlico.

Faulkner, W. (2009) Doing gender in engineering workplace cultures. Gender in/authencity and the in/visibility paradox. *Eng.Stud*, 1: 169–189.

Faulkner, W. (2014) Can women engineers be 'Real Engineers' and 'Real Women'? Gender in/authenticity in engineering. In E. Waltraud & I. Horwath (eds.) *Gender in Science and Technology*. Bielefeld: Transkript verlag. 187–204.

Foster, J. G., Rzhestsky, A., & Evans, J. A. (2015) Tradition and innovation in scientists' research strategies. *American Sociological Review*, 80(5): 875–908.

Frietsch, R., Haller, I., Funken-Vrohlings, M., & Grupp, H. (2009) Gender-specific patterns in patenting and publishing. *Research Policy*, 38: 590–599.

Gaughan, M., & Corley, E. A. (2010) Science faculty at US research universities: the impacts of university research center-affiliation and gender on industrial activities. *Technovation*, 30(3): 215–222.

Gebhard, M. (1947) *Lopen uupuneelle perheenemännälle*. Työtehotietoa-lehti 7–8/1947.

Ghiasi, G., Lariviere, V., & Sugimoto, C. R. (2015) On the compliance of women engineers with a gendered scientific system. *PLoS ONE*, 10(12): e0145931. doi: 10.1371/journal.pone.0145931.

Gianiodis, P. (2014) A framework for investigating university-based technology transfer and commercialization. In T. Baker & F. Welter (eds.) *The Routledge Companion to Entrepreneurship*. London: Routldge. 207–223.

Giuliani, E., Morrison, A., Pietrobelli, C., & Rabellotti, R. (2010) Who are the researchers that are collaborating with industry? An analysis of the wine sectors in Chile, South Africa and Italy. *Research Policy*, 39: 748–761.

Godin, B. (2008) *Innovation: the history of a category*. Working paper no 1. The project on the Intellectual History of Innovation. Montreal: INRS.

Godin, B. (2015) Models of innovation: why models of innovation are models, or what work is being done in calling them models? Project on the intellectual history of innovation. *Social Studies of Science*, 45(4): 570–596.

Gray, D. O., Tornatzky, L. G., & Rideout, E. (2014) Introduction. In L. G. Tornatzky & E. Rideout (eds.) *Innovation U 2.0: Reinventing University Roles in a Knowledge Economy*. Raleigh, NC: Southern Growth Policy Board

Greatest Engineering Achievements (2016) Household appliances history, part 3 – vacuums and fans. http://www.greatachievements.org/?id=3775. Retrieved 1.3.2016.

Harhoff, D. (2011) The role of patents and licenses in securing external finance for innovation. In D. B. Audretsch, O. Falk, S. Heblich, & A. Lederer (eds.) *Handbook of Research on Innovation and Entrepreneurship*. Cheltenham: Edvard Elgar. 55–73.

Hasse, C. (2008) *Draw the Line! Universities as Workplaces for Male and Female Researchers in Europe*. UPGEM project. Tartu: Tartu University Press.

Hauge, T. (2016) Academic capitalism in the age of globalization. *Higher Education Research & Development*, 35(4): 865–867.

Horizon (2013) Interview of the laureate of the 2013 FEBS/EMBO Women in Science Award, Dr. Almouzni. The EU research and innovation magazine. https://horizon-magazine.eu/. Retrieved 15.5.2016.

Hughes, T. B. (1999) Edison and electric light. In D. MacKenzie & J. Wajcman (eds.) *The Social Shaping of Technology*. Buckingham: Open University Press. 50–63.

Hunt, J., Garant, J.-P., Herman, H., & Munroe, D. J. (2013) Why are women underrepresented amongst patentees?. *Research Policy*, 42(3013): 831–842.

Joshi, A. (2014) By whom and when is women's expertise recognized? The interactive effects of gender and education in science and engineering teams. *Administrative Science Quarterly*, 59(2): 202–239.

Jung, T., & Ejermo, O. (2014) Demographic patterns and trends in patenting: gender, age and education of inventors. *Technological Forecasting & Social Change*, 86: 110–124.

Kanter, R. M. (2000) When a thousand flowers bloom: structural, collective and social conditions for innovation in organization. In R. Swedberg (ed.) *Entrepreneurship*. Oxford: Oxford University Press. 167–210.

Kanter, R. M. (2001) *Evolve! Succeeding in the Digital Culture of Tomorrow*. Boston, MA: Harvard Business School Press.

Kariv, D. (2013) *Female Entrepreneurship and the New Venture Creation: An international overview*. New York: Palgrave.

Klofsten, M., & Jones-Evans, D. (2000) Comparing academic entrepreneurship in Europe -The case of Sweden and Ireland. *Small Business Economics*, 14(4): 299–309.

Knorr-Cetina, K. (1999) *Epistemic Cultures: How the Sciences Make Knowledge*. Cambridge Mass.: Harvard University Press.

Kovalainen, A. (1995) *At the Margins of the Economy: Women's Self-Employment in Finland, 1960–1990*. Ashgate: Avebury.

Kugele, K. (2010) Analysis of women's participation in high-technology patenting. In S. Marlow and P. Wynarczyk (eds.), *Innovating Women: Contributions to Technological Advancement*, Contemporary Issues in Entrepreneurship Research, Vol. 1. Bingley, UK: Emerald Group Publishing. 123–151.

Ladd, A. L. (2014) Let's talk about sex, baby: gendered innovations in orthopaedic science. *Clin Orthop Relat Res*, 472: 793–795.

Landau, E. (2006) *The History of Everyday Life*. Minneapolis: Twenty-First Century Books.

Lariviere, V., Ni, C., Gingras, Y., Cronin, B., & Sugimoto, C. R. (2013) Global gender disparities in science. *Nature*, 12(504): 211–213.

Larkin, J. (1989) *The Reshaping of Everyday Life: 1790–1840*. New York: Harper Perennial.

Lee, N., & Motzkau, J. (2012) The biosocial event: responding to innovation in the life science. *Sociology*, 46(3): 426–441.

Lindberg, M., Danilda, I., & Torstensson, B.-M. (2012) Women Resource Centres – A Creative Knowledge Environment of Quadruple Helix. *Journal of Knowledge Economy*, 3: 36–52.

Lissoni, F., Llerena, P., McKelvey, M., & Sanditov, B. (2008) Academic patenting in Europe: new evidence from the KEINS database. *Research Evaluation*, 17(2): 87–102.

Longino, H. E. (1994) The fate of knowledge in social theories of science. In F. F. Schmitt (ed.) *Socializing Epistemology: The Social Dimensions of Knowledge*. Lanham: Rowman Littlefield.

Longino, H. E. (2002) *The Fate of Knowledge*. Princeton, N.J: Princeton University Press.

Lungeanu, A., & Norshir, S. C. (2015) The effects of diversity and network ties on innovations: the emergence of a new scientific field. *American Behavioral Scientist*, 59(5): 548–564.

Mavriplis, C., Heller, R., Beil, C., Dam, K., Yassinskaya, N., Shaw, M., & Sorensen, C. (2010) Mind the gap: women in STEM career breaks. *Journal of Technology, Management & Innovation*, 5(1): 140–151.

Maxwell, L. M. (2003) *Save Womens Lives: History of Washing Machines*. Eaton: Oldewash.

McKelvey, M., & Holmén, M. (2009) *Learning to Compete in European Universities: From Social Institution to Knowledge Business*. Cheltenham: Edward Elgar.

Meng, Y. (2016) Collaboration patterns and patenting: exploring gender distinctions. *Research Policy*, 45: 56–67.

Meng, Y., & Shapira, P. (2011) Women and patenting in nanotechnology: scale, scope and equity. In S. E. Cozzens & J. M. Wetmore (eds.) *Nanotechnology and the Challenges of Equity, Equality and Development.* New York: Springer.

Merton, R. K. (1968) The Matthew effect in science. The reward and communication systems of science are considered. *Science,* 159: 56–63.

Milli, J., Williams-Baron, E., Berlan, M., Xia, J., & Gault, B. (2016) *Equity in Innovation: Women Inventors and Patents.* IWPR C448. Washington: Institute for Women's Policy Research.

Mui, C., & Carroll, P. B. (2013) *The New Killer Apps: How Large Companies Can Out-Innovate Start-Ups.* New York: Cornerloft Press.

Murray, F., & Graham, L. (2007) Buying science and selling science: gender differences in the market for commercial science. *Industrial and Corporate Change,* 16(4): 657–689.

Murray, F., & Stern, S. (2014) Do formal intellectual property rights hinder the free flow of scientific knowledge? An empirical test of the anti-commons hypothesis. *Journal of Economic Behavior & Organization,* 63(4): 648–687.

Museums Victoria (2016) Item ST 26358 Washing Machine – Bendix, Automatic, circa 195. Museums Victoria Collections http://collections.museumvictoria.com.au/items/415562. Accessed 28 November 2016.

Nager, A., Hart, D. M., Ezell, S., & Atkinson, R. D. (2016) The demographics of innovation in the United States. ITIF. Web-version: http://www2.itif.org/2016-demographics-of-innovation.pdf?_ga=1.194345133.1854841563.1452803793. Retrieved 12.March 2016.

Nählinder, J. (2013) Understanding innovation in a municipal context: a conceptual discussion. *Innovation: Management, Policy & Practice,* 15(4): 315–325.

Nählinder, J., & Tillmar, M. (2013) Towards a gender-aware understanding of innovation: a three-dimensional route. *International Journal of Gender and Entrepreneurship,* 7(1): 66–86.

Nählinder, J., Tillmar, M., & Wigren-Kristoferson, C. (2012) Are Female and male entrepreneurs equally innovative? – Reducing the gender bias of operationalisations and industries studied. In S. Andersson, K. Berglund, E. Gunnarsson, & E. Sundin (eds.) *Promoting Innovation. Policies, Practices and Procedures.* Vinnova Report VR 2012:08. Stockholm: Vinnova.

Naldi, F., Luzi, D., Valente, A., & Vannini-Parenti, I. (2005) Scientific and technological performance by gender. In H. Moed, W. Glänzel, & U. Schmoch (eds.) *Handbook of Quantitative Science and Technology Research.* Netherlands: Springer. 299–314.

Nobel Prize (2016) Prizes and Laureates. www.nobel.org, accessed 15.1.2016.

Perkmann, M., & Walsh, K. (2008) Engaging the scholar: three types of academic consulting and their impact on universities and industry. *Research Policy,* 37(10): 1884–1891.

Perkmann, M., Tartarik, V., McKelvey, M., Autio, E., Broström, A., D'Este, P., Fini, R., Geunae, A., Grimaldi, R., Hughes, A., Krabel, S., Kitson, M., Llerena, P., Lissoni, P., Salter, A., & Sobrero, M. (2013) Academic engagement and commercialisation: a review of the literature on university–industry relations. *Research Policy*, 42(2): 423–442.

Peterson Mcintery, M. (2010) *Bara den inte blir rosa, genus design och consumption i ett svenskt industriprojekt.* Stockholm: Mara Förlag.

Polkowska, D. (2013) Women scientists in the leaking pipeline: barriers to the commercialisation of scientific knowledge by women. *Journal of Technology Management and Innovation*, 8(2): 156–165.

Poutanen, S., & Kovalainen, A. (2013) Gendering innovation process in an industrial plant – revisiting tokenism, gender and innovation. *International Journal of Gender and Entrepreneurship*, 5(3): 257–274.

Poutanen, S., & Kovalainen, A. (2016) Professionalism and entrepreneurialism. In M. Dent, I. Lynn Bourgeault, J.-L. Denis, & E. Kuhlmann (eds.) *The Routledge Companion to the Professions and Professionalism.* London and New York: Routledge.

Pulma, P. (1984) *Työtehoseuran kuusi vuosikymmentä 1924–1984.* Helsinki: Työtehoseuran julkaisuja 260.

Pursell, C. (1995) *The Machine in America: A Social History of Technology.* Baltimore: Johns Hopkins University Press.

Ranga, M., & Etzkowitz, H. (2010) Athena in the world of techne: the gender dimension of technology, innovation and entrepreneurship. *Journal of Technology, Management and Innovation*, 5(1): 1–12.

Ranga, M., Gupta, N., & Etzkowitz, H. (2012) Gender Effects in Research Funding. A review of the scientific discussion on the gender-specific aspects of the evaluation of funding proposals and the awarding of funding. Bonn: DFG, Deutsche Forschungsgemeinschaft.

Rommes, E., Bath, C., & Maass, S. (2012) Methods for intervention: gender analysis and feminist design of ICT. *Science, Technology & Human Values*, 37(6): 653–662.

Rosser, S. V. (2009) The gender gap in patenting. Is technology transfer a feminist issue?. *NWSA Journal*, 21(2): 65–84.

Rosser, S. V. (2012) *Breaking into the Lab: Engineering Progress for Women in Science.* New York: New York University Press.

Sandberg, S. (2013) *Lean In: Women, Work and the Will to lead.* New York: Alfred A. Knopf.

Sang, K. J. C., Dainty, A. R. J., & Ison, S. G. (2014) Gender in the UK architectural profession: (re)producing and challenging hegemonic masculinity. *Work, employment and society*, 28(4): 247–264.

Schiebinger, L. & Klinge, I. (2013) *Gendered Innovations. How Gender analysis Contributes to Research.* Directorate General for Research & Innovation.

European Comission. Accessed on 14th February 2016 from http://ec. europa.eu/research/science-society/genderedinnovations/index_en.cfm.

Schiebinger, L., & Schraudner, M. (2011) Interdisciplinary approaches to achieving gendered innovations in science, medicine and engineering. *Interdisciplinary Science Reviews*, 36(2): 154–167.

Schmidt, B. (2014) Women, research and universities: excellence without gender bias. In B. Thege, S. Popescu-Willigmann, R. Pioch, & S. Badri-Höher (eds.) *Paths to Career and Success for Women in Science. Findings from International Research*. Wiesbaden: Springer Verlag. 93–116.

Shen, H. (2013) Inequality quantified: mind the gender gap. *Nature*, 495: 22–24.

Simard, C., & Gammal, D. L. (2012) Solutions to recruit technical women. In *Anita Borg Institute Solutions Series, Anita Borg Institute for Women and Technology*. Palo Alto: Anita Borg Institute.

Sims, S. T., Stefanick, M. L., Kronenberg, F., Sahcedina, N. A., & Schiebinger, L. (2010) Gendered Innovations: a new approach for nursing science. *Biological Research for Nursing*, 12(2): 156–161.

Slater, D. (2014) *Who made that? Windshield Wiper*, New York Times Magazine, September 14, 2014, p. 22.

Smith-Lawton, H., Chapman, D., Wood, P., Barnes, T., & Romano, S. (2014) Entrepreneurial academics and regional innovation systems: the case of spin-offs from London's universities. *Environment and Planning C: Government and Policy*, 32: 341–359.

Sproule, A. (2000) *Thomas A. Edison, The World's Greatest Inventor*. Woodbridge, CT: Blackbirch Press Inc.

Statistics Finland (2016) *Employment Statistics*. www.stat.fi. Retrieved 12.6.2016.

Stephan, P. E., & El-Ganainy, A. (2007) The entrepreneurial puzzle: explaining the gender gap. *Journal of Technology Transfer*, 32: 475–487. doi: 10.1007/s10961-007-9033-3.

Stuart, T. E., & Ding, W. (2006) When do scientists become entrepreneurs?. *American Journal of Sociology*, 112(1): 97–144.

Sugimoto, C. R., Ni, C., West, J. D., & Larivière, V. (2015a) The academic advantage: gender disparities in patenting. *PLoS ONE*, 10(5): e0128000.

Sugimoto, C. R., Ni, C., & Lariviere, V. (2015b) On the relationship between gender disparities in scholarly communication and country-level development indicators. *Science and Public Policy*, 42(6): 789–810.

Technology Academy Finland (2016) Introduction and prizes. www.tat.fi. Retrieved 25.5.2016.

The Economist (2014) Innovation awards: and the winners are . . . The Economist 6.12.2014. www.theeconomist.com. Retrieved 12.10.2016.

Trask, B. S. (2014) *Women, Work and Globalization: Challenges and Opportunities*. London: Routledge.

Truss, C., Conway, E., d'Amato, A., Kelly, G., Monks, K., Hannon, E., & Flood, P. C. (2012) Knowledge work: gender-blind or gender-biased?. *Work Employment & Society*, 26(5): 735–754.

Työtehoseura (2016) Työtehoseuran historiaa. http://www.tts.fi/tts-1/ tts90vuotta. Retrieved 10.5.2016.

Valtonen, J. (2014) *The Serpent House – Protection and Reparation of Cultural-Historically Valuable Building*. Unpublished Master's thesis. Helsinki: University of Aalto.

Waltraud, E. (2014) Diffraction patterns? Shifting gender norms in biology and technology. In E. Waltraud & I. Horwath (eds.) *Gender in Science and Technology*. Bielefeld: Transkript verlag. 147–164.

Watt, H. M. G., Shapka, J. D., Morris, Z. A., Durik, A. M., Keating, D. P., & Eccles, J. S. (2012) Gendered motivational processes affecting high school mathematics participation, educational aspirations, and career plans: a comparison of samples from Australia, Canada, and the United States. *Developmental Psychology*, 48(6): 1594–1611.

Weber, S., Wiegel, C., & Busolt, U. (2014) The German business enterprise sector: career paths in Research and Development (R&D). In B. Thege, S. Popescu-Willigmann, R. Pioch, & S. Badri-Höher (eds.) *Paths to Career and Success for Women in Science. Findings from International Research*. Wiesbaden: Springer Verlag. 241–260.

Wennerås, C., & Wold, A. (1997) Nepotism and sexism in peer-review. *Nature*, 387: 341–343.

Whittington, K. B. (2011) Mothers of Inventions? Gender, motherhood and new dimensions of productivity in the science profession. *Work and Occupations*, 28(2011): 417–456.

Whittington, K. B., & Smith-Doerr, L. (2005) Gender and commercial science: women's patenting in the life sciences. *Journal of Technology Transfer*, 30: 355–370.

Whittington, K. B., & Smith-Doerr, L. (2008) Women inventors in context: disparities in patenting across academia and industry. *Gender and Society*, 22: 194–218.

Wihlman, T., Hoppe, M., Wihlman, U., & Sandmark, H. (2014) Employee-driven innovation in welfare services. *Nordic Journal of Working Life Studies*, 4(2): 159–180.

Yusof, N., Kamal, E. M., Kong-Seng, L., & Iranmanesh, M. (2014) Are innovations being created or adopted in the construction industry? Exploring innovation in the construction industry. *SAGE Open*, July-September 2014: 1–9.

Zmroczek, C. (1992) Dirty Linen. Women, class and washing machines, 1920s–1960s. *Women's Studies International Quarterly*, 15(2): 173–185.

CHAPTER 3

New Economy, Platform Economy and Gender

During the last two decades, the new economy has come to epitomize many things in society. Originally, the term 'the new economy' was used as a contrast to the 'old, industrial-driven economy', which refers to industrial production and material accumulation of goods and services tied to industrial goods (OECD 2004; also 2005). The original domain of the new economy and its rhetoric has since its early use widened further to embrace economic, territorial and intellectual areas. The term 'new economy' in general has been used to describe the aspects and sectors of economy that use, produce or are derived from new technologies.

The scope of the new economy has ranged from the technology bubble and dot-coms, powered by the rise of websites, Internet firms and the tech industry, to a more general concept including all types of immaterial and innovation-driven work, consumption and changing living conditions, both nationally and beyond borders. In the new economy, digital platforms have recently shaken both product markets (via, e.g. Apple and Google app stores) and labour markets (via, e.g. Uber, BlaBlaCars, Didi, TaskRabbit and Upwork) worldwide.

The de-coupling of technological development and industrial production shows how complex the developments of the economic drivers are: technological progress and technological developments and innovations are currently present in all new economy drivers and definitions, and increasingly, technological progress and innovations are entangled and elude economic interpretations.

© The Author(s) 2017
S. Poutanen, A. Kovalainen, *Gender and Innovation in the New Economy*, DOI 10.1057/978-1-137-52702-8_3

By definition, the new economy covers a variety of activities connected to technological developments and innovations in organic and dynamic ways. What differ are the understandings of the new economy and its scope. For some the new economy is indicative of rising inequality both between and within countries (Atkinson 2004; Piketty 2014) and that these developments are inevitably interrelated (e.g. Atkinson 2012; Perrons et al. 2007; Agnew 2001). These assumptions are often based on macro-economic analyses, but not derived from the technological innovations as such, nor from the gender perspectives (e.g. Balsamo 2014). The understanding of the new economy in relation to means of production directs the view towards the outcomes of production. Globalization and innovation do produce more wealth, but they also tend to increase social inequalities globally (Breznitz and Zysman 2013; Bair 2010; Baber 2001).

Where is gender located and how is it subjected to new positions and possibilities in the discussion surrounding the new economy? In the previous chapter we showed that despite the fact that innovations are becoming more and more crucial to contemporary economies, they are still highly gendered and gender-biased. In addition to gender aspects of innovations, there are other elements in the contemporary economies that are worth the analysis. In this chapter we argue that gendered work and labour – whether local or global – challenges both the alleged disentanglement of skills/capabilities from personhood and the assumption that a 'corrosion of character' (Sennett 1998; Sennett and Cobb 1972) will take place due to short-termism in the new economy. These two phenomena are discussed as key features of work in the new economy in the following section.

3.1   What is the New Economy?

Imagine the times in the 1960s when typewriters were mechanical and imagine a specific situation where an author runs out of ribbon in her typewriter, perhaps in the middle of a sentence and with a highly pressing timetable. At that time, almost all manual typewriters used half-inch-wide ribbons but the spools differed. Ribbons were usually sold on plastic spools and they were most often machine-dependent solutions because no standardized solution existed on the market. As a consequence, usually several kinds of ribbons with various types of spools would have been tried before the one that would fit the machine of our author could be found. If

the local bookstore did not store the machine specific ribbon her typewriter required, the only way to obtain the right kind of typewriter ribbon
was to order one by post. Ordering and receiving the right kind of ribbon
through a specialist shop or through a bookstore in the 1960s may have
taken several days.

What happened to the writing process during those days waiting for the
right kind of ribbon to arrive, apart from writing by hand? Most probably,
the writing and editing work stood still or would be done by hand. Today,
no such lack of ink and ribbon – or paper – would stop the author from
writing. The most usual interruptions in the contemporary world, not
counting power cuts, e-mails or social media, may occur due to the need
to recharge laptop batteries or upload a new version of the text processing
programme. In the 1970s, memory typewriters replaced repetitive retyp-
ing with mechanical typewriters in offices. In the 1980s, PCs and ink-jet
printers were introduced, followed by laser printers. The era of mechanical
typewriting was over in the 1980s. The computer did not wear out in a
similar fashion to the mechanical typewriter and correcting the text
became much easier, although machine and software versions changed
constantly. The mechanical was replaced by the electronic across the
whole of society, soon to be replaced by the digital. This happened
throughout society in all work tasks and jobs.

After the alliance of communication with the computer, and partly
through that, the birth of the Internet in the 1990s, start-ups in the
technology field and the growth of new companies riding the wave of
new technology was swift. Current well-established companies were
founded twenty years ago during the new technology boom. Of these
companies, Amazon.com was founded in 1994, Google was founded in
1998 and Wikipedia in 2001. The new social media corporations followed:
Facebook was founded in 2004, and Twitter in 2006. Old media corporations soon followed in transforming their business models as well
(Sundararajan 2016; Stephany 2015; Simon 2011a). The era of new
type of networked societies, with the growth of Internet-based innovations and businesses, was beginning to grow at the global scale.

In less than twenty years, technological inventions in the new economy
have multiplied and exploded, and the new types of interdependencies
between technological development and immaterial services have grown
significantly. The so-called 'start-up culture' began its rise in Silicon Valley
in the USA shortly after 2000 with new technology businesses and start-
ups, but it was rooted to a much earlier development and the close

collaboration that existed between universities and corporations (e.g. Kenney and Zysman 2015; Evans and Schmalensee 2016; Etzkowitz and Leydesdorff 2000). The global world has not been equal in this development: interestingly the technology boom that took place in India in the 1990s caught up much later and with a different flavour than in the USA, much due to outsourcing of work and innovation activities (Nadeem 2011), and was later followed by China with slightly differing pattern.

Technology in general has become more complex, entangled and layered, with design architecture systems ranging from hand-held devices to wearable digital technologies and the services around them, to the governance of smart homes for consumers and centralized cloud solutions and platform services for companies as part of their value creation. The seeds and the ingredients for feeding and scaling up the new economy as it is today were sown almost twenty years ago. But the new economy is also triggering other things than just technologically wired services and products. Technology is a great enabler and transformer, and it is therefore justifiable to ask what the gendered effects of it are and how gender, technology and innovations are intertwined and interrelated. The new economy as we know it today no longer concerns technology alone even if technology has irreversibly changed the ways our economies function. Instead of focusing on dealing with the scarcity of material production and costs of the products, the new economy has an 'abundance' of non-material products and digital technologies, and involves the increasing consumption of non-material products as well (e.g. Shaugnessy 2015; Dolgin 2012). As examples, the time structure and speed of the economic transfers have become instant, and thus new symbolic values are suffused into the new economy and have become global overnight (Evans and Schmalensee 2016).

One aspect often forgotten in the hyped discussion surrounding the new economy is the fact that the old economic structures have re-structured and transformed in concert with the new global economy. Enterprises are increasingly externalizing their activities, and that is taking a different form than previously: instead of subcontracting, more open forms of transactions are developing through new start-ups, open design and expanding ecosystems. Some authors argue that there is a discernible shift away from the overall importance of multinational corporations (e.g. Shaugnessy 2015; Mui and Carroll 2013). Whether or not the traditional enterprise model is breaking down or taking a different shape remains to be seen. The argument with the new economy innovations and businesses is that while the contemporary value of economy increasingly lies in the dematerialized assets, the governance of these assets becomes highly complex endeavour that requires fully new types of 'asset management' skills, starting from cultural capital.

Global economic competition, technological drivers and changes in industrial structures have challenged and put pressure on national labour market institutions, work-related benefits and other arrangements. These challenges and pressures have been analysed in detail nationally and globally (e.g. Kittur et al. 2013; Coyle 2011). The growth of atypical working contracts, the transformation of skills and knowledge, the polarization of the work force, and higher levels of unemployment as a result of globalization are current features of most labour markets (e.g. Felstiner 2011), and relate to gendered work in differing ways. Research in general agrees that the major disruptive force behind the current economic transformation is technological development (e.g. Goos et al. 2014) but there is disagreement on its consequences towards work, both in terms of the contents of work and in terms of various working arrangements (e.g. Adams and Deakin 2014; Manning 2013). For example, in the case of knowledge services clusters are increasingly important form of the local dense networks of work, and bring in one additional aspect to innovations. According to several authors (Manning 2013; Cooke 2002), knowledge service clusters are also globally geographic concentrations of lower-cost skills that serve global demand for increasingly commoditized knowledge services.

Burning societal issues, which range from the mismatch between education and the skills and capabilities needed for working life, to the high level of recognition of business opportunities by the working age population in combination with a low number of newly established businesses, require a detailed analysis of the new economy in relation to economic fluctuation and global economies.

Not all of the changes in the research landscape are a direct consequence of external economic shocks. The economic downturn during the first decade of the twenty-first century has more or less globally introduced new types of demands for and pressures on national innovation systems and their related subsystems. Measures required due to economic austerity and economic pressures, such as cutbacks in budgets, also speed up changes in the innovation infrastructure landscape. As noted, in the USA, one tenth of the labour force participate in a platform-mediated "gig economy" in some capacity (EY 2015). The phenomenon of 'platform economy' remains less prevalent in many European countries but is on the rise, boosted not only by examples of global platform companies, such as Uber, but also by the common policy of a single market of EU for the digital age and the recent rise in crowd employment and its new related forms throughout Europe (e.g. Eurofound 2016). The idea of 'gig economy' as understood precarious, temporal and short-term work has earlier concerned only those with less education and qualifications than today is the case. By changing the value creation and organization of existing work, new platforms also enable new forms of work to emerge. More broadly, the contents and arrangements of work are changing in concert with the digital development of the economy.

The idea of the new economy has thus transformed the economy from being a site-based industrial activity directed by technology to a more inclusive cloud based way of thinking involving digital platforms and this changes the services provided, the work and the workers involved. Technology is replacing people in many contemporary occupations, and this trend has immense gendered effects. If, for example, 'personhood' becomes a more important selling point and the appealing 'vehicle' in the new economy than it has earlier been, this may define genders anew.

Technology has accelerated because of technological development, and the new economy, based on technology, has enabled world-wide globalization by facilitating the rapid spread of technical progress, programmes and applications, and with these, the assumption of a global universal culture. The typical features of the first phase of globalization through technology included the open system of trade, rapid information flows and the spread of technology on the terms and in the image of Western society. This is also the most usual definition of globalization in relation to technology development and rise of innovation.

However, defining the new economy and innovation only through technological progress and a rather one-dimensional view of technology sets aside a more complex understanding of how innovations and inventions are highly embedded in societies and localities, and how they are interconnected with societal developments. The competitive advantage for the nation-states of the new economy is increasingly based on innovation of both tangible and intangible elements and the generation of new business models, and not only in technologies but also in social areas. In this the global interdependencies become visible when countries and contexts are analysed. For example, for India, the high-tech outsourcing industry has grown immensely in the last ten years and is a matter of considerable pride for the country, with a high number of global corporations situated in India. The view of the global corporations is different: the outsourcing industry is viewed as a low-cost, often low-skill sector (Nadeem 2011; Shehzad 2011). Despite the fact that workers use the digital tools of the information economy, they do not necessarily work on complex or technologically innovative tasks. Through lively ethnographic detail and subtle analysis of interviews with workers, managers, and employers, Nadeem demonstrates the culturally transformative power of globalization and its effects on the lives of the individuals at its edges.

The literature very often defines the first stage of technological globalization – the prerequisite for the new economy – as being based on early web-based designs and developments (e.g. Simon 2011b). Web-based design started a type of technological globalization that was not related to standardized production or mechanical products, but more to communication, computers and the Internet. In the Internet world, web-based design rather quickly included development and a move from personal websites to other forms of presence in the Internet such as blogging (Flew 2007). With this, the globalization idea grew with the uniform ideology of growth in standardized technologies, spreading with a rather universal culture. At the same time as it created multiple site development opportunities, it also allowed the standardization of a universal culture through algorithms and because of similarities in the communication cultures.

All innovations are highly gendered in many ways. The definition of an innovation we adopt in this book is wide and does not only take into consideration the traditional definition of the commercialization of an invention through market demands (Braunerhjelm 2012). Indeed, as markets are not a single phenomenon but several, and as markets can be both created and constructed the distinction between inventions, innovation and imitation may not be so clear-cut but rather blurred. As we will show later in the book through examples, innovations can remain very local and restricted.

The current view on the complex role of technology and technological innovations in global development and economic growth is reflected in the concepts 'globalization 2.0' and 'globalization 3.0'. Both versions are in contrast to globalization 1.0 and refer to the individualized solutions and differentiated developments, but also to new forms of non-Western modernity that have become increasingly important as part of overall globalization, and the constant interdependence of economies. Globally, the advent of this new era has been hastened by the fiscal and financial crisis in Europe and in the USA, and political instability in many of the Arab states (e.g. Li 2012). The range of complicating issues indeed extends from financial to political crises, conveying the scope of the dynamism that relates to the phases of the economies in question.

The growth in the 'diversity of globalization' may help raise productivity and wages at the nation-state level through innovation, entrepreneurship, markets access and trade channels. Today the globalization of innovation and R&D in companies literally takes them, if needed, everywhere: thus a laboratory in India may have software from Finland, centrifuges from Japan, the laboratory ventilation system from the UK and automated biological testing kits and systems from the USA. But the new economy does not only consist of the most developed technologies and their global access and implementations. It requires an economy that draws from and is strongly built on – not only smart copying or replicating – but on innovations and inventions people create, which make a difference to the existing world. Most often, the new economy still needs a workforce, somebody to work in the laboratory and use the automated machinery, despite the growing robotization and automatization (e.g. Frey and Osborne 2013).

Evidenced by a multitude of research (Sennett 2006; Vallas 2011; Edgell and Vogl 2013; Breznitz and Zysman 2013), the trends of digitalization, robotization and globalization have changed, and continue to profoundly change, both production and services, and more generally, the idea of work and working life. The spread of ideas, economic forces and technologies alike has become an irreversible process. The consequences of these changes are currently visible in contemporary working life and in the qualifications required and the demand for both low and highly skilled occupations and work, and are much researched (e.g. Goos et al. 2014; Kalleberg 2011; Oesch and Rodriguez Menes 2011; Acemoglu and Autor 2011).

The transformations brought in by global economic and societal trends create new platforms for services with new jobs and while transforming old ones, often with at an intensifying pace and with exponential value creation potential at the global scale. Currently c. 50% of the world's highest valued brands are associated with digital platforms (EY 2015). These new platforms for their part enable new forms of work, and they transform and change old forms of gendered work, and create new jobs and businesses (Mandl et al. 2015). The polarization of the workforce according to capabilities and skills in technological and digital fields is a well-documented fact. The individualization of the skills and capabilities in the so-called 'third spirit of capitalism' (see Boltanski and Chiapello 2006; Hardin 2014; Harvey 2005) has importantly reshaped the gendered characteristics of corporate citizens in recent decades, with renewed emphasis on corporate femininities and masculinities e.g. Poutanen and Kovalainen (2016; Sandberg 2013). How do these capabilities and skills relate to gender of the new innovative worker, requires further analysis.

One essential element of innovation is the intellectual property rights (IPR) which in most countries have become very protective (e.g. Carayannis et al. 2015). Research and development yields product innovations and is thus the key to economic growth. It is interesting that at the national level strong IPR protection relates positively to the innovation capability of the country, even if the connection is complex. It may sound very counterintuitive, but as all innovation investments include a risk for a company or for a start-up, if the IPR system is nationally weak, it fails to protect the rights of the company and thus reduces the interest in making investments in R&D and innovations of the companies located in that country. If the companies have the ability to protect their innovation capabilities, they will invest elsewhere. If the IPR protection at the national level is weak, fewer people accept the risk of copying or theft involved in innovation investments, which then nationally slows down the innovation rate. Still, globalization has globally made national borders porous and the flow of ideas and networks has never followed nation-state borders.

But not all innovations in the new economy are built around the global, technologically cutting edge innovations. One of the aspects brought up in the gender literature is the lack of the user perspective, and more deeply, the understanding of the processes in which the technology becomes usable and practical. The lack of gender perspective and more profound gender aspect in technology development and design is given as one reason why many start-ups are 'not changing the world': the teams who construct the products and design the applications of technologies are simply not representative of the world, or even of the Western population (e.g. Leonard 2002; Oldenziel 2004).

Innovation literature shows, among other things, that the gender composition of R&D teams has an effect on the nature of innovations, as it directs the ideas and the ways the ideas are worked through and put forward in the team and/or company (Poutanen and Kovalainen 2013; Diaz-Garcia and Welter 2013; Sastre 2016). The results resonate with the vast literature on the composition of problem solving groups in organizations and the impact of diversity on innovation. According to most results, with several kinds of empirical research settings, diverse groups outperform homogenous groups by a significant margin (e.g. Hong and Page 2004; Lin 2015). Additionally, some research has even shown that when a problem appears to be difficult, diversity trumps individual abilities (e.g. Kotiranta et al. 2010). For any company that wishes to stand out with continuous innovation in order to stay competitive, diversity becomes an important issue.

In Chapter 4, we will discuss in detail the global nature of innovations in relation to services and we will also examine social innovations. Many of the contemporary innovations are complex entanglements of social and technical innovations with contextual features and processes. For example, some innovations have been developed for saving unpaid household work. Innovations are a complex process overall and research shows that many factors affect it. A nationally built intellectual property system alone does not sufficiently produce innovation (e.g. Lewis 2008), as innovations are no longer nationally bound. Still, nation-states do invest in innovation policies and activities much more strongly than earlier, despite the fact that the new economy and its innovations do not recognize national boundaries but are global from the very beginning. As argued, the life of an innovation, from its research and development process, to the final marketable product or service across frontiers does not know national boundaries (Gass 2008). During their lifecycle, even global innovations generate jobs and new innovations that benefit the state, and most importantly, they generate networks of knowledge, capabilities and elements of success when fruitful.

In the long run, for any economy, growth in productivity has been argued as the single most important economic indicator. For these reasons, the growth of the new economy has not been seen to be as prominent as it was assumed it would be. Ranging from the shop-floor level to management, the ways in which the work and production processes are organized anew with the digitalization, for example, requires time and work. The desire to perform a job well is also shaped by the interaction with technology (Thursfield 2015; Berner 2008), and new technologies complicate old working habits. On the other hand, any economic growth will bring forward some part of the new economy as growth in general makes room for new investments such as digitalization. This will transform 'old' industries and 'old' innovations alike, especially if we think of work and production as a heterogeneous assemblage of socio-material practices, where digitalized production and human beings interact.

In order to remain competitive, highly successful firms not only appear to continuously develop a stronger alignment across their strategies, structures, and processes, but they also seek ways of understanding how those elements fit together (Miles and Snow 2003). Miles and Snow (2003) note, that at the firm-level efficient organizations constantly modify and refine the mechanisms by which they achieve their goals. Previous research indeed shows that innovative enterprises, new firms, as well as large established businesses willing to exploit new business opportunities though entrepreneurial activity, all play an important role in the complex network and process of technological change, amplifying the renewal of industries (e.g. Acs and Audretsch 2005).

The productivity paradox, brought forward by Robert Solow who once famously noted that 'you can see the computer age everywhere but in the productivity statistics' (Solow 1987), describes the problem of measurement for productivity in recognizing the transformation. Information technology alone does not increase productivity. When companies reorganize their production around technology and digitalization, in order to make the new technology work for its benefit, then statistic measures need to recognize the change. However, this is not a straightforward linear development, nor is it without gendered effects. One of the arguments most often put forward for closer linkages between innovating companies and higher education institutions (universities and research institutes) is the stimulus and influence of the academic spill-over on companies (Van Beers et al. 2008).

If innovation is to be understood as part of the work in organizations, new businesses, universities and in R&D in general and if it is regarded as an irreplaceable part of the work, then innovation work should also be seen as situated practice, and as an enactment of performance, rather than being understood as an institutionally organized action that only takes place in pre-determined settings (e.g. Orlikowski 2008). The situated practice-perspective and enactment-perspective both have a different type of explanatory power when analysing innovations and the gendering of innovations. The two perspectives also relate to gender in different ways. The practice perspective questions the explanatory power of the macro-level analysis concerning various aspects of innovations. Even though usability, spread and usefulness may be the acid tests for innovations, the practical aspects to do with inventions and birth of innovations is a highly processual and micro-level phenomenon, which relates to gender and gendering of innovations in several interesting ways.

3.2   CHANGING RELATIONSHIP BETWEEN GENDER, WORK AND CAPITAL

The new economy is full of promise: the simultaneous presence of technological developments and platforms will boost new start-ups and new, economically viable businesses, transforming existing industries and enabling new business models to appear also in services. At the same time the new economy boosts overall productivity and job creation. In this, the interdependencies of the global economies on the national and regional scale are clear. The demands for better productivity at the national level direct attention not to technologies as such, but to the ability to innovate and direct such activities through policy instruments.

The need to focus on a more responsive science and technology policy follows global trends for greater efficiency, accountability and control in research systems (e.g. Power 1997; Elzinga 2004; Webster 2013), but acknowledges greater needs for innovations. These changes have been discussed in terms of 'national systems of innovation' (e.g. Nelson 1993), 'research systems in transition' (Ziman 1994), or even 'the post-modern research system' (Rip and Van Der Meulen 1996), and they all emphasize the role of knowledge in the economy and society. It has been argued that the 'knowledge complex', the set of formal and informal institutions that support and mediate knowledge exchange has drastically and irreversibly changed (Murray et al. 2009; Murray and Stern 2014) and a new generation of knowledge interdependence (e.g. Powell and Giannella 2010) has been brought about by innovation, companies and universities alike. The new institutions for their part change and blur the boundaries in knowledge production, mediation and innovation activities.

The overall growth based on the current global flux of ideas, innovations and exchange of knowledge and products, call it Industrial revolution 4.0, Globalization 3.0 or Globalization X.0, has increased the importance of regional and local hubs as the melting pots of the new economy. Globalization connects the new economy to specific regions, where economic dynamism and blending processes that often result in novelty are to be found. In fact, globalization highlights regional differences as it enables them to build strengths in contrast to other regions. The importance of cities and urban areas as social and economic melting pots, knowledge creators, centres of innovation, and drivers of social and economic change emphasize the differentiated development within states and for the future of regions.

The dynamics of creative and innovative jobs, which are considered as indicators for the new economy, occurs in urban areas and cities and regions with economic vitality. Dynamics requires capital, which for its part is no longer national or local, but global, adding one dimension of complexity to interdependence. Economic vibrancy can be analyzed in traditional measures, such as the number of patents issued, investments in research and development, and in the number of IPOs (initial public stock offerings) (e.g. EY 2014b). Furthermore, investments in green solutions in businesses, the number of entrepreneurs and new start-ups in the region, job dynamics, job churning, and the number of scientists and engineers in the workforce may all contribute to indicating the innovation capacity of the region. The global effects are actualized in different ways at local levels.

According to recent research in the UK, it is far more often the creative occupations rather than small creative firms that innovate (Lee and Rodriguez-Pose 2014; Schonfield 2011), indicating that the design-based occupations – many of which consist of self-employed persons – employ more people than design-based industries altogether (Vallance 2015). Many of these occupations are one-person companies which work in highly networked economy, with no intentions or possiblities to high growth. This result confirms that creative occupations play an important role in innovation and in innovating networks, and that their innovating is often done through contractual work, not through paid employment in industry. The new dynamics of economy requires capital, which is, as stated, global, adding global interdependence.

At the national economies' level, there is not just one way to track down and measure the new economy or innovation. There are several nation-level indicators that are attached to the innovations and inventions, and more indirectly, to the new economy. Often three attributes are attached to the new economy: global, computer-based, and innovative attributes. These three attributes stem from the definition given by the OECD in 2004 of the term the 'new economy': the new economy was seen as describing aspects or sectors of an economy that produce or intensely use innovative or new technologies, applying to industries where people depend on computers, telecommunications and the Internet to produce, sell and distribute goods and services (OECD 2004). The most typical measures and attributes at the state level highlight the traditional role of R&D, namely the amount of active patenting and the type of IPR the nation states have. But there are other measures that are associated at the nation-state level with the new economy and its various features. These nation-level measures often include measures such as the share of technology producing industries of all industries, the existence and vibrancy of the venture capital industry in the region or in the state, the number of entrepreneurs and new businesses, the percentage of knowledge jobs out of all the jobs available, the numbers of new business start-ups and so on. (e.g. Atkinson 2012).

While some of the nation-level measures are easily spotted as gendered measures, such as the number of entrepreneurs and share of new business owners, as well as knowledge based jobs, some of the measures may seem gender-less, such as measures of risk funding or the vibrancy of the venture capital industry. This is not, however, the case. The venture capital business is geared towards risk funding for entrepreneurs, seed funding for start-ups and early stage investments. The business idea in venture capital is that of raising money and making money for investors.

The gender differences in gaining access to external equity capital are evident. The gender gap with respect to external equity funding was analysed using German data on new ventures founded between 2005 and 2009 by Lins and Lutz (2016). The data covered over 3,100 new ventures, and the analyses showed that female entrepreneurs received less venture capital in Germany for new ventures than male entrepreneurs. In the German data, the effect was particularly strong among entrepreneurs with university degrees and entrepreneurial projects with high research and development activities (Lins and Lutz 2016).

The venture capital sector is a highly male-dominated industry with only a few women present as partners or senior advisers in venture capital firms globally. A piece of research carried out by Babson College called the Diana Report was an initiative working to increase the number of women entrepreneurs, and it found that the number of women partners in VC firms decreased from 10% in 1999 to 6% in 2014 (Greene et al. 2001; Brush et al. 2009; Brush 2014). Recent research reports show that the disparities in venture capital funding relate to the business plan but also to the lack of technical education, especially so in the technology wired environment (Tinkler et al. 2015).

The venture capital markets are overwhelmingly most developed in USA and the majority of companies and capital are US based. The majority of the world's annual venture capital investments is US-based venture capital activity (Statista 2015), and European venture capital activity accounts for only minority. In the USA, two areas arise above the others: Silicon Valley (Bay Area) and Boston-NY area (MIT area). Globally, new countries have also risen among the investors: the large emerging economies – China and India have doubled their proportion globally. As a consequence, China has surpassed Europe as the world's second-largest market for venture capital. Equally, India has surpassed Israel at the global level being the fourth largest market for venture capital (EY 2014a).

The earning logic for venture capital is that by investing in the developmental phase of the company a VC firm will be making money on the future valuations and the future potential of the company it is investing in. Of the 100 largest venture capital firms in the USA in 2015, 40 did not have any women among the venture capitalists involved (SFGate 2015). Much as a result of this disparity, some new venture capital firms have been launched to offer financial capital to the underserved market of women-led companies, focusing on funding women owned businesses. Venture capital has also shifted from seed stage funding to later stage venture development (Greene et al. 2001; Tinkler et al. 2015). Research has suggested the intricate ways in which gender is involved in the venture capital decision making process and how it remains gender imbalanced (e.g. Tinkler et al. 2015). The effects of social networks may also play a crucial role, and to combat this, some venture capital firms focusing on women entrepreneurs have networked with business angels and early stage investors in tech.

When innovations require new investments for start-up or newly emerging companies, global activity does not mean equality in terms of available capital. Not only have the European governments started to invest in complementary funding, but also, there is a difference in the capital investments globally: European and Asian venture capital investments focus on the consumer services sector, in the USA (as well as in Canada and Israel) the main emphasis of the venture capital firms is on information technology companies.

Against this large backdrop, new funding instruments have arisen with the shrinking globe. Crowdsourcing has become one of the major disruptors of the financial markets in view of access to capital and ways of obtaining it. Crowdsourcing as a new way of outsourcing tasks, resources and activities works also in the financial sector. The idea that innovations are critical to knowledge based economic growth is already well-established in classical economics, and is widely accepted, as is the idea that efficient innovation policies boost national economies in most developed economies. Innovation policies have been highlighted in recent years in national and supranational research policies and the emphasis on the outcomes, innovations and economic returns on immaterial investments have become more prominent than earlier (Rider et al. 2013).

In Europe, according to the EY barometer from 2014, private funding does not replace public money: the second most important source of funding for new innovative companies and new innovations are government programmes, direct investments or aid. Government funding may partly consist of research based funding, as in the not-so well-known case for Google. Google's early funding was from the National Science Foundation (NSF), which funded the invention of a successful algorithm. Thus, the state also has an important role to play in the USA, where the overwhelming majority of investments of the private types are made. Hence, it is understandable that the solely women owned or women led venture capital firms reside in the USA where the capital markets are most developed. Within USA, two main regions, Silicon Valley/Bay Area and New York/Boston serve as hotbeds for venture capital firms. Both the Bay area and New York have also grown in importance globally (EY 2014a).

Research evidence shows that also in funding, and especially in venture funding, diverse leadership teams that are comprised of women and men produce better results than all male teams. Women entrepreneurs report in several research reports that if they have a choice, they would prefer to have women investors and additional top female talent on their boards of directors and advisory councils (e.g. Kotiranta et al. 2010). Questions about the intersection of gender and venture capital financing were largely unexamined until the early 2000s (Lins and Lutz 2016). Greene et al. (2001) found in their exploratory study, which utilized longitudinal data to track US venture capital investments by proportion, stage, industry and gender, that there is an apparent gender gap in the funding. The recent growth in women owned venture capital businesses is based on a meticulous market analysis which found an untapped market where a high number of women and their gender-diverse teams were found starting scalable businesses with innovation potential in record numbers, and on a finding that in USA less than 10% of venture capital is invested in management teams that include women (Women's VC Fund 2016). The new economy is not restricted to measures that most often focus on the economy alone, but extends to educational and cultural fields as well. Measuring cultural and non-material elements such as the vibrancy of culture, for example, is notoriously difficult. Hence, the focus on the economic and socio-economic proxies enables some general level comparisons, even if the culture as well as the region is often left out in such comparisons. The Information Technology & Innovation Foundation has measured US society's ability to create and support the new economy at the state level with clusters of measures and indicators (Atkinson and Nager 2014). The measures used differentiate the states from each other, and also within the states, differences over time vary and seem to be growing for various reasons. The key areas that are assumed to best capture what is new about the new economy are all interrelated and assumed to benchmark economic dynamism. The key areas relate to knowledge-based jobs, the globalization of the economy and its dynamism, the growth of the digital economy and the innovation capacity of the state/region (e.g. World Bank 2016).

Knowledge work and jobs relate directly to the value added capacities of education in society in general, and value is mediated by occupations and industrial sectors. Hence, knowledge job indicators may include factors such as IT occupations in non-IT sectors, the education level of the workforce, the employment share in high-value-added manufacturing and the share of the workforce employed in managerial, professional and technical occupations (e.g. Webster 2013). Knowledge work and jobs include increasing variety of jobs that require highly specialized skills, creative capabilities and tacit knowledge. Jobs such as web-designers, editors, media specialists, ADs and copywriters no longer contain specific tasks only, but may require specializations that come with the knowledge acquired both through education and experience.

Knowledge work is not a twenty-first century phenomenon. As a concept, knowledge work ranges from the 1980s through the huge expansion of knowledge employment in the financial services sector of the advanced economies (McDowell 1997; also, 2008a), to the extent, that urban theorist Manuel Castells (1996) argued that technological innovation had created a compression of time and space with the constant global flow of money and information. But, as Robins has remarked, globalization also triggered a new dynamics of re-localization (Robins 1991; also

McDowell 1997). The national economies are no longer the boundaries for recruitments as the labour markets for scientists and engineers, for example, are global (e.g. Taylor 2010; Sweet and Meiksins 2012).

In these trends of knowledge and work changes, gender has become ingrained in many ways into the new jobs and new types of knowledge based work tasks that relate to innovations and the new economy surprisingly quickly. Some new jobs have become classified as 'women's work' or 'men's work' very quickly. One such job is coding. Already in school, the coding is often seen as boys' activity. Encouraging girls to coding takes extra efforts, such as the cases in Chapter 4 show.

Technology has a social shaping function for any type of knowledge work, giving it specific characteristics, yet, it is elusive in its many forms. In explaining social shaping terms such as 'embeddedness' (Granovetter 1985; Zelizer 1987) and 'embodiment' (Bourdieu 1984) are often used. In statistics, knowledge jobs are defined as clear categories: usually including jobs requiring at least a two-year-degree in college, and most often also including managerial, professional expert or technical aspects of the jobs. In the new economy, these jobs can also be found in fields other than IT sectors, due to the continuing digital transformation of even the most traditional work and industrial sectors. Similarly, the educated workforce is crucial for the full functioning of the new economy. This education most often is in science, technical, engineering and mathematics (STEM) fields, as discussed earlier in Chapter 2. STEM fields most often lead to occupations where expertise and skills together co-produce innovations. But innovations are not derived from education alone.

Looking into the organizational processes surrounding knowledge based jobs, gender and innovation, only a few studies can be found where a comparison has been extended over time and across countries and for both genders. The studies that analyse the experiences of men and women in knowledge intensive work show that gender differences prevail. Truss et al. (2012) show in a study from the UK and Ireland involving 498 male and female knowledge workers in knowledge intensive companies that despite equal levels of qualification and experience, women are more likely to be in lower status and less secure knowledge related jobs in these firms. For women the jobs available, even with the same type of work description in the companies featured less variety and autonomy than the jobs that men had. It appeared that despite comparable levels of knowledge, women were less likely to be in a position to translate their skills into the innovative work behaviour necessary for career advancement

(Truss et al. 2012). These empirical findings suggest that the experiences of women and women's participation in knowledge processes within knowledge intensive companies differ fundamentally from men's, thus providing a clear rationale to look more deeply into the actual processes in which the inequalities and gendered positioning takes place.

On the other hand, gender becomes embedded in knowledge work in a variety of ways and intersects with a variety of factors, such as work culture and history, and competitiveness in the work place. A recent ethnographic study of how the resistance to teamwork became prevalent in knowledge workers revealed that it was mostly shaped by laboratory workers' individualistic interactions with technology, the laboratory layout and the workers' desire for personal task-related autonomy and individual responsibility rather than team-based accountability (Thursfield 2015).

When looking on a more general level at the indicators of the new economy, knowledge jobs and education needed for knowledge work, a crucial question is, whether women have equal opportunities in STEM (Science, technical, engineering and mathematics) education and training leading to knowledge occupations and jobs in the new economy. The gender division in technical education is highly gendered globally (e.g. Ridgeway 2011; Ridgeway and Correll 2004; Moghadam 2000), and this extends beyond technical training into the training for manufacturing jobs as well. In computer science and in engineering, the percentage share of women's Ph.D. degrees is c. 20–25% and similarly 20–25% of the STEM jobs in the USA (NCWGE 2012). In the literature, women are often described as potential and an 'untapped opportunity' for the state and for the improvement of national competitiveness (e.g. Beede et al. 2011).

Globally, the numbers of women in the science, technology and innovation fields are low in the world's leading economies, including the USA (Schmidt 2014; Schiebinger 2008). In Ding et al. (2006) large study focusing on STEM academics with and without patents, the researchers found no difference between men's and women's scientific impact, when measured through standard measures. However, the gap in patenting between genders was found. This finding lead the researchers ask for the reasons for such a large gender difference in patenting. One plausible explanation is that men and women do qualitatively different kinds of research. Interviews showed that networks influenced on patenting behaviour, which is consistent with research findings on academic entrepreneurship. Most academic women had lesser contacts with industry than their male colleagues. Furthermore, of those interviewed, more women than men were concerned that pursuing commercial opportunities might hinder their university careers (Ding et al. 2006).

In relation to the efforts being made to give women greater access to science and technology education, some research shows negative results, particularly in the areas of engineering, physics and computer science. Women remain under-represented in degree programmes for these fields— less than 30% in most countries, despite clear intentions to raise their participation in knowledge based jobs (e.g. Beede et al. 2011). When analysing knowledge jobs and work in STEM fields in research universities, the picture is even more gendered concerning the division of labour for faculties in STEM fields (Carrigan et al. 2011; see also Howard 2016; Jardins 2010; Hewlett 2007).

Carrigan et al. (2011) examined the connections between time allocation and work efficiency for a STEM faculty in a National Study of postsecondary faculties carried out in 2004 with a weighted sample of over 13,000 participants, women and men. Throughout the data, researchers found a gendered division of labour and time use, and when the number of women went up, the division was more even. Similar results of the importance of a critical mass of any gender, with smaller samples and in different research settings have been found in the male dominated industries (Poutanen and Kovalainen 2013). Results also lend empirical support to theories that argue that critical mass of women (or men in female dominated workplace) attainment positively impacts equity in resource distribution and time allocation. According to a recent study of 40 years of development (1970–2010) of the women's share in STEM fields of higher education, the overall trend is one of increased access. Although the women's share has increased in STEM fields of study, women have still less access to engineering than natural science (Wyer et al. 2013; Carrigan et al. 2011; Ramirez and Kwak 2015). This does for its part explain the scarcity of women among inventors. Patenting is a crucial part of innovation work, and knowledge oriented jobs. A recent study on collaboration, gender and patenting with US data showed that increased collaboration with industry would significantly increase the probability of patenting for female academic scientists (Meng 2016).

The argument that having only a very few women who obtain patents hurts scientific innovation, technological development and national competitiveness, is based on the idea of gender equality and equal access to resources and rewards. This thinking extends to the idea that the predominance of men in patenting may mean that innovations that are useful for a broader population may not be developed, because women have invented many technologies for the home and for caretaking (Rosser 2009). However, because science is gendered, so are inventions and innovations for their parts. The main aspects of innovation do not necessarily gain from equality directly as such (Blake and Hanson 2005) but from freedom and opportunities for the "creation and exploitation of new ideas" (Kanter 2000: 168; Acker 2004).

All too often innovations and patents referred to in the literature are technological innovations rather than services and social innovations (e.g. Arthur 2007). In innovation research, product innovations are the most usual way to consider innovations: that is the creation of new products or adding new elements to old products, which can be measured either through patenting or through new product announcements (e.g. Feldman 2000). An alternative measure of innovation is process innovation, which is more difficult to quantify: process innovations concern the ways that firms organize and change in their activities. More precisely process innovations are often defined as those that "incorporate new technology into the method of production" (Feldman 2000: 374).

Another distinction often made is the difference between radical and incremental innovation, referring to the degree of 'newness'. Incremental innovations relate to small improvements or changes, and radical innovation refers to new products or categories of products, requiring new types of competencies (e.g. Blake and Hanson 2005). Both radical and incremental innovation can occur in product and process innovations. The two types of innovations, product versus process innovation on the one hand, and the two qualities or sorts of innovations, incremental versus radical innovation on the other hand add several new dimensions to the thinking surrounding innovation. This raises the important question of how gender relates to these very different types of innovation concepts.

Much of the innovation research has focused on technology in a rather narrow sense, thus in practice relating to product and process innovations is mostly taken to mean some form of technological change (e.g. Kanter 2000; Simmie et al. 2002; Blake and Hanson 2005). The importance of an innovation is defined by its capability to provide advantage in the markets to its producer. Less attention in the literature has been given to employees in the new economy and the continuous maintenance and enhancement of their skills and capabilities demanded by their employers. More attention has been directed at the building of a culture of commitment through several means, ranging from dress codes and providing snacks to spatial arrangements in the working environment and defining and sharing internal collective values (Baldry et al. 2007). The occupational community, in which "the worlds of work and non-work are closely inter-dependent, each world permeating and affecting each other" (Salaman 1974: 45), is highly important in the innovative new economy.

The contemporary concept of innovation refers most often to certain kinds of economic activities and very often excludes other sorts or less valuable economic activities, such as social innovations. Rethinking innovation requires positioning it in relation to gender and changing economic regimes. Banking on the potentiality of an innovation is one new aspect in the global market place and patenting field. It is thus the markets that ultimately define the 'value' of the innovation. Yet the markets are not perfect and cannot always set a 'price' for an innovation: the more radical the innovation, the more difficult it is to set the price. Decoupling the innovative activity with export-based (Blake and Hanson 2005) or economic exchange value theory adds a new dimension to the analysis of innovations.

When we move away from the gendered divisions in education, organizations, work, jobs and occupations, and delve into a more dynamic, constructed ideology of gender and work, it is the enactment of gender that becomes part of the embodied and performed gender. An ethnographic study in five research intensive university labs of female and male chemistry graduate students showed how women and men are socialized to "do masculinity" differently in scientific contexts in the laboratory where collaborative elements in discussions, team work, and research development settings were on the agenda (Hirschfield 2015). Using results from nine months of observation and 40 semi-structured interviews Hirschfield notes that the existing and prior linkages between men, science, and academia may allow male, but not female graduate students their freedom in establishing their claim on the scientific authority in the group. Hirschfield's (2015) research shows that male graduate students were able to express themselves in more varied and complex ways than their female peers. In contrast, female graduate students may feel more pressure to conform to the strict norms of competition that are associated with traditional masculinity.

Software engineering is a good example of an ideal career in knowledge based work, where variety is high in the contents of the work, but the marketable skills are high as well, creating a situation where young skilled workers are the human embodiment of the employing organization and its values, but only to the extent that the work is central to any person's life. Software engineering is also a very gendered occupation, with strong divisions between men and women in the field, their promotional opportunities and the ways in which the leaky pipeline in academia differs from other disciplines.

Additional aspect, further discussed in the Chapter 4, is the aspect of recruitments, and new types of gender biases that emerge with the new economy. As an example, the global multinational companies invest significantly in their recruiting infrastructures and possibilities to hire through campus recruitments, for example. Research has revealed some continuously existing gender biases in these increasingly invested hiring practices (e.g. Simard and Gammal 2012) that are worth revisiting anew when the new economy is being investigated further.

It is beyond dispute that technological progress has a direct connection with the economic development. Over the past 50 years, Western economies have increasingly shifted from physical goods production to an economy of innovations and an economy of knowledge. With this shift, much of the work has also drastically changed and the jobs have transformed from manual labour to the utilization of human capital. The rhetoric of the knowledge-based, innovation economies and societies has invaded our way of thinking about society and value creation in the Western world, the global north, and increasingly also globally (e.g. Centeno and Cohen 2010). Innovations become entangled with existing cultures and part of the societal tapestry, but also with existing industries and businesses.

Currently digitalization and the increasing use of complex computer algorithms are shaping and transforming industries and businesses anew globally. In fact, one might assume that with the increasing digitalization the relevance of locality evaporates. This is not, however, necessarily the case. If we leave aside the possibility that the locality may be a variable in an algorithm leading to local songs appearing on global music play lists and so on, there is also a more real-life meaning to locality that is present even in the global digital world. The surplus value of knowledge-based products is increasingly based on the alliance of innovation and design, but also on the new ways of producing products and services, and many of them are produced locally.

It has been estimated that for every, directly innovation based high-tech job, such as a software engineer hired at Google in Silicon Valley, there are according to Moretti's research (2012) five other types of job openings in the close-by area, most often in services, but both in skilled professions, such as lawyers and nurses, and in occupations, such as hairdressers and waiters. The multiplying effect means that even small increases in spending through consumption can lead to larger increases in economic output. The multiplying effect, which means the ability of higher salaried jobs to create other jobs, and the ways in which jobs are interconnected, is based not only on direct salary differences and the purchase power of workers' salaries, but also on more indirect effects of culture and society (see, e.g. EY 2014b; Clark et al. 2006). The multiplying effect makes the global economy locally based, but not necessarily local in the traditional sense of the word (e.g. Sassen 1996; Carr 2009; Caraway 2007; Beck 2000; World Economic Forum and Boston Consulting Group 2016). One of the measures for the new economy is the openness of society to migrants, the ability to use the skills and capabilities of migrants, and the ability to attract highly skilled migrants (Atkinson and Nager 2014; Atkinson and Andes 2010). Global competition may in fact enhance the global transition to a skilled and educated labour force with intensifying speed. In the USA, foreign-born and foreign-educated scientists and engineers have been more highly represented than US-born scientists among authors of the most-cited scientific papers and among inventors holding highly cited patents (Stephan and Levin 2001).

Due to the interconnected nature of the economy, the lingering effects of the high tech innovation economy relate to all consumption and investments, so, in practice this mean to all jobs with salaries or compensation. The innovation sector of the knowledge economy and jobs it provides have often been shown by economists as the strongest sectors for producing jobs in other sectors. The innovation sector has a three times larger effect than the manufacturing sector in creating new jobs (Moretti 2012). The differences between regions grow, as the ripple effects of the innovation sector can be global but are also often local and regional: services are consumed within a relatively close distance to the workplace and home. The rise of artisanal, local and small-scale production, local food shops, craftsmanship (Sennett 2006) and services interestingly follows the economic boost of innovation, in vibrant regions and local spaces. Locally rooted craftsmanship usually has no scalability, which has become the most important virtue required for any innovation to become profitable. Even if the business idea of some form of craftsmanship is based on the locality, scalability can be found in the ways of making the sales nationwide, such as farmers' markets, for example.

The world economy is, however, shaping whole nations, much due to fluctuating finances, recessions and economic expansions, but perhaps even more importantly, due to shifts and transformations in industrial production and services due to innovation. The external threat of globalization has actually become one important driver for productivity gains in national economies. But globalization as such does not make innovations happen, as they only come possible through human activities and interactions locally in communities, vividly pointed out in a classical study by Jacobs (1961).

Jacobs underlined the interaction within communities, the social fabric of the community which creates sufficient meeting places, agoras which enable platforms for innovation. Those platforms or 'hubs' require knowledge based on education provided by universities. The relationship between universities and innovations is complex but undisputable, with effects that radiate beyond the actual physical location of the university. The complexity of interwoven education system and innovation system linkages is a well-known established feature of entrepreneurial and innovation hubs from Shanghai to Silicon Valley (e.g. Saxenian 1994). In countries where a national innovation system is highly interconnected with the private sector, the relations between university led research and private sector R&D are well developed and interconnected (Rogers 2016; Laestadius and Rickne 2012; Rickne et al. 2012).

The knowledge-based economy –thinking has come to mean that creativity, innovation and entrepreneurship are as such desirable and that these elements should become the capabilities of the whole population, not only the well-educated part of the population. We may ask where these capabilities will become actualized? Are they well-suited qualifications for all jobs and occupations? The individual capabilities lead to a creative class of professionals and innovative behaviour that leads to new business models and eventually to new products (e.g. Shaugnessy 2016; Florida 2014; Berkun 2010). The key hub for such a transformation is education, most often delivered by universities. In universities the emphasis has traditionally been on developing and promoting individual capabilities and skills, and not on developing innovations as the main 'product' of the university. University-borne innovations have often been presented to be more of as happy by-products of the education and research when successful, and not the targeted aim of the education system as such. The multiple roles of universities have however become under scrutiny due to increasing financial pressures and reciprocal relations vis-a-vis businesses and industries.

The multiple roles and tasks of the universities are present in the current research landscape, perhaps more than ever (e.g. Poutanen and Kovalainen 2010; Laestadius and Rickne 2012). Such discourses usually ask the question of whether universities stand for knowledge or for innovations, alternatively, they may discuss how universities provide both wellbeing for citizens and the sustainable renewal of society, if not through knowledge then through innovation. Most often these two aspects of understanding the purpose of universities and the knowledge they impart are found in the best universities incubating innovations, but not always. The structural policies, complex networks and top-down support systems consisting of the universities' innovation services (formerly often technology transfer offices, TTOs) are needed within the universities, to back up regional and local economic innovations, with job-creation policies (Clarysse et al. 2007).

The rise in the number of academic spin-offs seems to be dependent on several contextual factors that range from the ownership of intellectual property rights, to the commercialization of research and to shrinking university basic funding. The Bayh-Dole Act in 1980 adopted in the USA was followed by a rise in the number of formal commercial knowledge transfers from US universities to companies. There were three main mechanisms through which such activities were accelerated in universities: licensing, joint ventures in research and university based start-ups (Link et al. 2007). The change both in the types and in the ways universities participate in financing the spin-offs has become more varied and adjusted to the local triple helix – that is, the connections between businesses, governance and university – activities.

The Bayh-Dole legislation in the USA has led to similar types of legislation on intellectual property rights in several countries, which allows for a university to receive the intellectual property rights, which is often seen as first step towards the professionalization of the TTOs (Siegel and Wright 2014). While the funding mechanisms vary, one of the common features of European university spin-offs is their reliance on public funding in the initial funding stage, in preference to finance from venture capitalists, business angels or industrial partners (Autio et al. 2014; Siegel and Wright 2014).

The creation of enabling state-level, regional and local mechanisms on the one hand, and the mobilization of human and cultural resources, on the other hand, should be the aim of new innovation policies (e.g. Geels 2014). They also have several implications for understanding what the economic activities are, what are counted as innovations, and how these are measured. If the science and innovation policies are geared towards hunting for the next best new innovation, they may fail to find it due to the strict narrowing down of both the scope and the type of innovations they are targeting.

3.3   PLATFORM ECONOMY, GIG ECONOMY AND SHARING ECONOMY

Digital platforms have shaken both product markets (via, e.g. Apple and Google app stores) and labour markets (via, e.g. Uber and Upwork) worldwide. In the USA, one tenth of the labour force participate in a platform-mediated "gig economy" in some capacity (EY 2015; Seppälä et al. 2014). The phenomenon remains surprisingly less prevalent in many innovation-driven economies, but is on the rise, boosted in Europe by the EU drive for a common policy of a single market for the digital age, as well as the recent rise in crowd employment and its related forms throughout Europe (Eurofound 2016). By changing the value creation and organization of existing work, platforms also enable new forms of work to emerge and new types of innovations to appear, even outside the existing organizations.

More widely, digitalization embraces almost every aspect of contemporary work and ranges from local care services to highly specialized, technology-led and cloud-based services. Thus, work contents and arrangements change in concert with digital developments. The insecurities in societal and economic developments have given rise to new forms of employment. Increased flexibility is currently required of employers and employees. The characterization and theorization of new work and new employment forms are still very much under way (e.g. Felstiner 2011; Rouvinen and Kenney 2015; Rubery 2015).

Throughout this chapter, we have discussed the new phenomena of 'platform economy', 'gig economy', 'sharing economy' and other new ways of organizing activities for the digitalized economies. How to define these variants and how to differentiate them from each other? All the new 'forms of economies' discussed here are some derivatives of the digital development and the possibilities it enables. For that reason, the discussion is sometimes elusive and sometimes polarized, without strong cases of evidence. This is partly due to the currently insufficient measures of the new phenomena, partly a question related to the strong contextual nature of these activities (e.g. Brinkley 2016). Some most common definitional characteristics of each of these new phenomena are necessary, in order to position these vis-a-vis each other.

The platform economy is emerging as a technologically driven and technology dependent development that shapes work, institutions, organizations, value chains and business models. According to recent analyses (Seppälä et al. 2014; Seppälä and Mattila 2016), the platform economy can be defined and understood as (i) resettling the contemporary industry boundaries and architectures, (ii) changing the logic of value creation and value capture, (iii) reconfiguring labour, and (iv) re-positioning competitive advantage globally as an enabler and a disruptor. Given all conditions above it is clear that the platform economy is highly contingent on the existing economic and societal structures. The platform economy is also a set of functions and modes operating across the economy, mostly 'boundary agnostic' of conventional industries (Seppälä et al. 2014; also Seppälä and Mattila 2016). Technological solutions such as technology platforms do not dictate the outcomes of the use of platforms, because national economies and sectors are historically structured and differ in their functions and activities (Kenney and Zysman 2015), but they do spread globally available and scalable solutions.

'The gig economy' is elusive to strict definitions. While the technological developments enable the appearance of 'gig work' more often than earlier, the technology does not determine its existence alone. Occasional on-line working, based on the skills and capabilities or professional skills may not count as employment or specific form of self-employment but more as additional job. In Hathaway's study, where he looks at the changes in the employment in non-employee firms in San Francisco in two sectors, in transport and in accommodation in 2009–2013, major changes were visible (Hathaway 2015; Hathaway and Muro 2016). In this study, clear indication of the gig economy was found but the elusive nature of the phenomenon comes clear in other studies with different data sets, such as by Katz and Kruger, who analysed the tax return forms and formation of salaries from different sources in the same area (Katz and Krueger 2016). When looking at the evolving process by way of aggregate self-employment statistics, the platform economies are not easily traceable (e.g. Zumbrun and Sussman 2015).

Gig economy is related to but differs from the 'sharing economy'. Both are used invariably but with slightly differing emphasis. As the insecurity has increased at the mature labour markets, with digitalization the work has become easier to organize in fractional ways. Yet, in most countries, the effects of the 'gig economy' are not visible in the increasing numbers of short-termism at the labour markets. The robust indicators and measurements for the 'gig economy' are still lacking. It can be assumed that some part of the growth in self-employment is one indicator of the growth of the 'gig economy', but is it alone a sufficient measurement? The evidence for a strong shift into the nonstandard or contingent forms of work is still lacking to some extent, at least in USA and in UK (e.g. Bernhardt 2014). Short-term work has become more available, and platforms have enabled more widely available short-term job markets to develop.

Technological platforms, for their part, irreversibly change the landscape for industries and organizations alike. As national boundaries have become more permeable and local business ties have loosened, with platforms the transnational competition has challenged labour markets, shifted firm level tasks and triggered new types of entrepreneurship. The growth of atypical working contracts, the transformation of skills and knowledge, polarization of the work force, and higher levels of unemployment as a result of globalization are current features of the new economy labour markets (e.g. Felstiner 2011).

Evidenced by a multitude of research (Sennett 2006; Vallas 2011; Edgell and Vogl 2013; Kenney and Zysman 2015) digitalization, robotization and globalization profoundly change production and services, and more generally, the idea of work. The spread of ideas, economic forces and technologies alike has become an irreversible process. The consequences of these changes are currently visible in contemporary working life and in the qualifications needed and demand for low and highly skilled occupations and work (e.g. ibid, Kalleberg 2011; Oesch and Rodriguez Menes 2011; Acemoglu and Autor 2011). Entrepreneurial renewal of the economy takes place through existing firms, the public sector and through new firm creation. Firm-level regeneration of skills and capabilities is necessary for the development of industries, but also for the emergence of new types of jobs and occupations.

Forecasts estimate that the current pace of technology development will, in industrialized countries, erase 36% or more of current occupations in 10–20 years (Pajarinen and Rouvinen 2014; Frey and Osborne 2013). The topical questions thus are: to what extent does education take into account the changing needs for qualifications, skills and capabilities of future labour? Also, in relation to this, in what ways will education respond to future challenges by ensuring the relevance of education to changing skills demands, as well as changing labour markets and economic environments, and to calls for different capabilities?

Currently c. 50% of the world's highest valued brands are associated with digital platforms (EY 2015). These new platforms enable new forms of work and transform old forms of work, while creating new jobs and businesses (Yeung 2003; Mandl et al. 2015). While new businesses account for a smaller share of new jobs, more jobs are created by firms that grow. There is strong evidence that a few rapidly growing companies create a disproportionally large share of new jobs both in Europe and in the USA. The polarization of the workforce is a well-documented fact both in Finland as it is elsewhere. How do firms and the public sector – facing the challenges briefly outlined earlier – adapt to and adopt new working modes and develop relevant skills and capabilities? Global trends such as digitalization transform work as they enable various platforms to develop and take over value chains. New dimensions and skills requirements are added to old jobs (Sennett 2006), but also old jobs and competencies are more thoroughly transformed, some disappearing and some becoming more fragmented in terms of their forms and working hours, among other things (e.g. Vallas 2011; Rubery 2015). It is still open as to how and to what extent these major global transformations translate into local changes.

Some of the business models of the new economy, exemplified with companies such as Uber, Airbnb, Taskrabbit and several other more unknown brands, have brought forward new labels. These labels such as the 'platform economy', 'gig economy', 'peer economy' and 'sharing economy' are used invariably in the popular press and to some extent also in the economic media. None of the labels explain in a detailed manner, how these business models function, who they bring in profit for and what types of complex and modern interdependencies their business models require in order to function (Kenney and Zysman 2015). The new corporations of the new economy, such as Airbnb or Uber, work on the basis of the vibrancy of the locality, strength of the culture and functions of society itself. The digital business model of the platform economy is in fact inseparable from society and culture at large: e.g. in most probability nobody would use Airbnb or other rental apps company to rent a room in a small remote village in Northern Sweden with no particular attractions on-site, while the odds to get a spare room rented out through Airbnb in London or New York are much higher, due to the societal and cultural attractiveness of the cities.

The risks and rewards of being entrepreneurial in the platform economy vary. If we use the spare room renting as an example, the risk taking of an individual in renovating his or her flat or house so as to be representable and attractive for potential customers of platform economy -company becomes palpable if one compares small remote rural town and vibrant big city. A small, rather worn-out room in a flat in London or New York would attract viewers and renters on Airbnb pages or similar websites if the location and price match expectations. A house or a room in Northern Scandinavian remote town would not draw similar attention. The business opportunities that can be built on the 'locality' depend entirely on other factors, such as cultural interests. The dependence on the location varies. There are differences between 'locals' which can be translated into the global markets. In the rent seeking business model based on the location, for example London is globally attractive, and brings in revenue through the cultural vibrancy of the city, while a small village in Northern Sweden may not be, and does not exist on that map.

The business model of these new economy companies is based on global and local layers of culture but follows in general the rent seeking model: instead of franchising a flat or a house with standardized furniture, or a hotel-type of solution, the business models of the new economy bank on the 'authenticity', and 'experience of locality'. Still, the business model assumes some form of global knowledge and shared understanding of the attractiveness, which can only be built by other than business means. This 'authenticity' has its own cultural codes and parameters in order to work as a business. By this we mean that it is more interesting and culturally almost self-evident to travel to London, Paris or Rome, instead of a remote place with no known global cultural, historical or contemporary attractions, if one is to consume the 'organic urban vibrancy', as Jacobs (1961) expressed. When the attraction to 'authenticity' and 'local living experience' becomes part of the business model, it is evident that the value added by the rental company is in fact earned through the investments the renting person makes in own housing, own cultural capital and own standard of living. The relationships between investment, cost, rent and price need to be thought seriously anew and will be defined in a new way.

Many of the business models of the new economy extend the risk taking to the individuals themselves, for example through the complexity of home-owners' own investments in renovating the flat or room in hope of a return on their investment through rentals. The renting host will not necessarily be able to rent without the platform the company offers, yet, he or she is expected to follow the rules of the company – as peasants do. What is interesting is that in the process, the home becomes commodified – a place where normal living becomes somewhat regulated due to constant monitoring and web-site reports by renters.

The oldest of the 'sharing economy' businesses was company called Couchsurfing. The Couchsurfing website opened in 2003 with the opportunity for a homeowner to open up his or her home to be used by other people (www.couchsurfing.com). Couchsurfing net achieved a registered user figure of 10 million people. Couchsurfing was one of the first successful new economy businesses and it is still open. A newer sharing economy business has been manifested in Zipcar which rents out cars on an hourly basis. Besides resources, labour may also be hired out, as in Uber, Upwork and TaskRabitt, some to mention. Several nationally differing national and even local adaptations of the labour hiring platforms exist, too many to be listed here.

Platforms are shared resources which work as base for business activities and innovations. The platform economy transforms the scaling of business activities into network effects. With network effects, the platform economy changes the concept of innovative services and products into scalable issues. The idea of the very different network effects and varieties is demonstrated by comparing two such companies – Uber and Etsy – two platforms that are both technology enabled, in fact, technology created, and yet, while one of the platforms is most often introduced and labelled as innovative and entrepreneurial, the other is often introduced as the "hottest hot". There are differences between these two though, as one of them is closer to a new type of subcontracting and gig economy relations in terms of the "freedom to work" and "need to work for the money" type of entrepreneur, while the other is closer to traditional self-employment and small-scale entrepreneurship. As economic models, they are hybrids and consist of many layers, partly traditional and partly digital activities.

One vibrant example of the new economy is Etsy which was founded in 2005 as an on-line selling place for handmade and vintage items, arts and crafts supplies, handicrafts and designed clothes, jewellery and gifts. In slightly over 10 years of active on-line selling, Etsy has continued to grow. Etsy is currently among the more established and prominent platforms for creative and innovative handicrafts and designs, and there are several others. Etsy and websites like Etsy – all platform sellers of handmade products – also interestingly represent the new forms of the gendered economy and a new type of stability in gender relations in the new gendered economy. One of the gendered features of Etsy that relates to platform is the social element with bulletin boards, discussion forums, offline meetings and exchange of knowledge offline.

Etsy has currently c. 1.5 million sellers, of which 86% are women, often with customer loyalty. Etsy has been able to create a new global and gendered economy of local design and handicrafts. If we take range of c. 54 million registered users, c. 25 million shoppers and more than 1.5

million sellers who sold goods through the Etsy on-line ecosystem in 2015 (www.etsy.com), both the earning and the consumption opportunities within Etsy are truly scalable, even if the variation of different artisanal cultures is not as high as it could be (Close 2016).

3.3.1   Case: Accountant Susan Makes Jewellery and Sells it Globally – is it a Hobby or is it a Job?

Susan is an accountant in a medium-sized accounting company, and her career has been stable. The stability has given Susan the opportunity to also enjoy other things in life apart from work. Recently Susan's hobby, which is jewellery making from sterling silver with professional polymer clay has taken up more and more of her free time. Therefore, she has decided to set up a business in her home, to make jewellery. She works with oven-hardening clay with pure pigments and uses both sterling silver clay and solid sterling silver. This means that she has installed an extra oven for the polymer clay and mostly works with a jeweller's saw, soldering kit, hammer and anvil, and sometimes with a blowtorch. Working with her own designs has brought her many compliments and she has been able to develop a "brand of her own", with colourful designs. Susan's designs have been very successful in local markets.

When Susan decided that her handicraft jewellery would need a proper and permanent marketplace instead of a shared market stall in her home-town's annual literary festivals and other such places, she started to look for alternatives for reaching a larger customer group. On-line selling came up through a work place contact. However, starting an on-line shop of her own was out of question, as she worked full time as an accountant. At the time when she heard about Etsy and decided to sign up with them, she really did not know what Etsy might do to her small-scale handicraft design production. Indeed, she did not know anything about the platform economy, "crowd-based capitalism" or global market places enabled by the new structure of digital platforms. For her, Etsy was first and foremost an Internet community of like-minded talented and creative persons, rather than a global show-case and potential global shop for her products and designs.

But within two months, Susan started to receive orders, first in the USA and soon internationally. Susan's company pages at Etsy became highly visited, and the comments she received through the Etsy webpage applauded her design and use of colours. Currently Susan ships her jewellery worldwide on a regular basis. The shipping methods she uses are secure and parcels can be tracked and insured. The stability in the ways she handles all her orders make her production trustworthy and reliable – she has been able to build a strong brand with the help of Etsy. She intends not to give up her accountant work but has agreed with her employer to be able to work fewer hours in order to pursue her handicraft business which has become global almost overnight and clearly without a proper business plan.

Indeed, through virtual community building, Etsy is an example of those digital platforms that enable the sellers of manually produced and individually designed goods and handicrafts to build virtual communities and teams, and collectively draw on other sellers' expertise in their business development. In this sense, Etsy is not only an Internet-enabled global marketplace and platform but also a virtual community through which "crowd-based" capital accumulation and new forms of earning emerge. This new form of capitalism enables new innovations. The new way is the way of the "crowd" – be it then seed funders or buyers to become the owner of the markets (Sundararajan 2016; Dourish and Bell 2011; Benioff and Alder 2009).

Going back to Susan and her business: After receiving orders from Europe, Susan no longer thought of her jewellery design and preparation work as a mere hobby. The money she receives from her orders in fact barely covers the costs of the design and production, as she uses almost all her spare time in trying out new ideas with materials, and designing new models which never see the light of day, as she says. If she would count those hours as billable working time, it would result in a calculation where no financial reward could be gained from the hours she puts in into her business. For Etsy, she pays commissions based on sales, but not a monthly fee. Etsy is a 'benefit corporation' (so called B-Corp), which relies on co-endeavouring and commissions by sale and by display (www.etsy.com). The buyers of the products pay taxes and shipping fees. For crafters, Etsy offers a rather easy global site for sales and a site for possible social, personal or business transformation.

3.3.2 Case: New and Old Economy Companies and the Ways Gender is Entangled in their Business Models

Two very concrete examples that manifest the contrast between the old and the new economy are the birth of Instagram and the death of the Eastman Kodak, more widely known as Kodak. Both of these companies, Instagram and Eastman Kodak deal with imaging products, pictures, photographic materials, images and imaging fields, but with very different types of logic, history and development (Brynjolfsson et al. 2014). The differences between the two are rather striking, despite the fact that photos are at the core of both companies. Instagram, the on-line mobile photo-sharing and video-sharing application based on technology platform, is based on the network effects of digital materials, while Eastman Kodak was mainly based on paper and photographic film, cameras and later on printers and ink. At its peak, Kodak employed c. 145,000 people with billions of US dollars as an asset base. Instagram, for its part, was established in 2010 and designed by people who built a digital product, which is free to use. The owners sold Instagram after a year's development to Facebook in 2012 (Business Insider 2014). Instagram is shared by millions of people, currently approximately with over 200 million users.

Instagram belongs to a different era than Eastman Kodak, which was established 1880. By 1976, close to its 100-year birthday, Eastman Kodak accounted for 90% of film and 85% of the camera sales in the USA. The years following were to change the development of the company, leading to a situation where, in 2012, Kodak filed for bankruptcy (The Economist 2012). The brand that was rated once among the world's five most valuable brands was too slow in its business decisions and moves to transform its key business activities in the wake of digital photography and for new types of consumers. Kodak extended photography from studios into family life and the home. Indeed, Kodak for years built its business with a heavy reliance on female consumers in their roles as mothers and daughters who documented family life, gatherings and events throughout their lives. This strategy was sustained for many years forming the core of Kodak's advertising campaigns. Dr. Munir points out, that the Kodak ad campaigns specifically emphasized how women were supposed to document and archive family vacations, events and family history. It was indeed through this gendering of family photography and campaigning for the "Kodak moments" and the "Kodak girl" that created entrenched, familial, everyday concepts (Munir 2012; Cooper 2003; West 2000) which were based on a gendered division of family life which no longer held true as digital trends and ideas of women being the main customers changed.

The original idea in the introduction of the "Kodak Girl" in the 1880s was to sell more cameras, but the image was also an active subject, instead of being a passive, camera-selling female. From early on, especially in the 1920s, the image of the "Kodak Girl" in ads was active: she was knowledgeable with her camera, she travelled widely, and she took pictures on holiday and sent them home as postcards (Cooper 2003). The "Kodak Girl" became the symbol of the home snapshot picture taking. The importance of home as the key subject of Kodak ads grew over time, and it has been argued that through the taking of snapshots Kodak taught people to see their experiences as objects of nostalgia and as a chain of festive events such as birthdays, parties, family gatherings and holidays (West 2000). The slogans accompanying the "Kodak Girl" included, "Save your happy memories with a Kodak", (Beardi 2015), suggesting the social desirability of everyday life in pictures. Pictures were taken to be shown to others and documenting everyday life and success in it. The gendered imagery of the home was present in the photos, even if women were targeted as active buyers and users of cameras. The intended gift, as one Kodak ad put it, with a picture of a gift box, wrapped in beautiful paper and ribbons, with a simple text on the card: "Open this first" was intended as a gift for family gatherings and events, as well as documentation of the success of the family life.

Contrasting Instagram to Kodak would not do justice to either of them. Instead, similarities can be found in the ways the Instagram offers a platform for advertisements through its clients' demographics. The overwhelming majority of the 150 million people on Instagram are under the age of 25, making it a lucrative platform for consumer marketing. Breaking down the demographics and usage behaviour, Instagram skews heavily towards women: 68% of Instagram users are women (Business Insider 2014) and Instagram does not strongly deviate from other social new media which focus on images, pictures and videos such as Pinterest, Vine and Snapchat (e.g. Shaugnessy 2015). Women presumably form the majority of the users of these social media channels and apps.

The analyses of the failure of Kodak are many, usually arguing for too 'small' thinking by the Kodak management (e.g. Mui and Carroll 2013) and strive to continue business as usual despite the changes. From the gender point of view contrasting Kodak with its rival Fujifilm is interesting as it brings in gender into the focus of the old and new economy corporations yet in additional way.

Fujifilm faced similar problems as Kodak with the rise of digital cameras that brought about drastic declines in demand for film. Fujifilm did develop the first digital camera, but instead of technical cameras opted for developing its laboratories and new products. In 2006, the company changed its name from Fuji Photo Film to the current name, Fujifilm and launched its own women's cosmetics. From then on, Fujifilm turned its strategy towards new feminine consumer markets with new products based on earlier inventions and innovations while the company still was on film business. Fuji's laboratories had developed technologies to prevent the colour fading in films, taking place in oxidation process by ultraviolet rays, in order to preserve vivid colours in printed photographs. These technologies were transferred – through new innovations – into the growing business field, cosmetics, attracting Asian markets and women buyers.

It may be much too straightforward to argue that Kodak failed because the top management was unable to see and sense two fundamental changes: the change in the gender positions within the family and in society and the role of digital imaging and photography at large. The argument however holds true as it reflects Kodak's failure to adapt to the new technology, despite the great resources it had technically, and failure to adapt to the changing consumer patterns. The inability to connect to changing customers and changing technology failed Kodak in the end. Inability to think deeply what it means when the gendering of the products becomes possibility (as for Fujifilm) and widening the product range to respond to changing customer needs were among the biggest reasons for Kodak's failure (e.g. Poutanen and Kovalainen 2013). The gendering can take many forms, as this case shows.

3.4   GENDERING THE PLATFORM ECONOMY

This chapter has analysed the ways the new economy and new platforms transform the earlier ideas of the innovations, and also of gender and innovations. We can ask what exactly then changes in the development of the new digital, networked economy? It would be easy to reply with the sole focus on a new business model which moves from individual businesses into sharing economy models, with early predecessors in eBay (Brynjolfsson and McAfee 2014; Brynjolfsson et al. 2014). Business models are, however, part and parcel of a larger transformation that concerns industries as well as whole economic and financial sectors. The transformations are deeply effectual as the changes concern the whole logic of the business and give rise to new types of businesses, business models and logic, but also lead to a new type of consumer culture and its unpredictability, as evident in the presented cases.

The economists that try to locate the new economy and its growth in the key macro-economic indicators argue that the new economy has not actually changed much when macro-economic indicators are scrutinized. The enduring relationships that govern macroeconomic processes, that is, the relationships between growth, unemployment, inflation, and budgets, have not changed in any fundamental way. But perhaps the measuring tools, such as the often used total factor productivity measure are too robust to register the influences of the networks and connections and other more immaterial effects discussed above. To capture the impact of innovations and the ways the new economy affects the organization of work, innovation and public and private relations may require other means of measurement.

That the innovations are bound to their time and location is shown in research, for example, by Gordon (2016) who demonstrates that the kind of life-altering scale of innovations that took place between 1870 and 1970 cannot be repeated as such (Gordon 2016). The life-altering scale can be exemplified in numerous inventions which have had an effect on other events and activities. One such example of an irreversible influence triggered by innovation is the invention of electric elevator by von Siemens in 1880, which immediately allowed the way of building houses to change: elevators allowed housing to extend vertically instead of horizontally, and as a result the electric elevator fully changed the construction industry and the land use planning in towns and cities.

The macro-level effects of the new economy in terms of general changes in the contents of people's work and job descriptions are visible in the increase of structural unemployment. The risks of unemployment in the new economy have become more dispersed than earlier, ranging through all age groups and qualification ranks (e.g. Lippman 2008; Wilton 2011; Daguerre 2014). Structural unemployment differs by nature from so-called cyclical unemployment, which is connected to a recession and is often time-bound. Frictional unemployment, the 'normal' turnover between jobs does not reflect the new economy either. Structural unemployment at the macro-level can reflect the transition to the new economy in cases when a mismatch occurs between the skills employers want and skills of the workforce.

Almost fifteen years ago, in 2003, The Economist announced that "In at least one sense, America's new economy is well and truly dead," and continued, "in another sense the new economy is very much alive and kicking," (The Economist 2003). In the USA investments and improvements in productivity based on new information technology were clearly growing in the early 2000s, to the levels which in 2002 were the highest since World War II. Since then, the acceleration in productivity has grown, not only in the USA but also globally in many industrialized countries, but with instability and not necessarily following a particularly clear trend. Within economics, the saying is that the productivity debate is currently surrounded by a metaphorical thick statistical fog. At the national level, statistics may give an unsubstantiated figure of growth for many countries, while not all improvements and investments in the quality of products and services are visible in the productivity statistics. The value-added indicator is still the strongest indicator of the presence of the new economy within the old industrial sectors. But does it work for the new forms of platform economy, is one crucial question that needs to be addressed, when capturing the gendered effects.

The downside of the platform economy in general, whether we discuss the 'gig economy', 'shared economy' or 'digital platforms', the most usual labels used, is that it puts the emphasis on new types of short term job contracts, which are neither freelancing nor employee positions. The platform economy refers to a more inclusive and wider idea of the connections and enabling 'agora' for businesses and people. The sharing economy brings up the wider issue of consumerism, which can be – to a tiny degree – alleviated through sharing things and commodities. It is, however, justified to ask whether these phenomena are in any specific way new, or whether they are manifestations or modifications of existing economic models, but in new format due to digitalization and electronic means of delivery, for example. Sharing is thus not 'sharing' in the actual meaning of the word ("what is mine is yours", "you're free to use what is mine") but an exchange of commodities, which range from individual cultural capital, to networks, materials and labour. What is different is the extension of commodities to intellectual and human capital, including your own networks, families and relations. We may ask, whether the complexity of the stated new economy and its relations shortly outlined above position women's innovative potential into a new frame. Furthermore, and in relation to that, what are the gendered effects of the new economy in its different forms?

In terms of the jobs available in the new economy, and the gendering of such jobs, the picture of whether the new economy strengthens the gender segregated labour market patterns or whether it enables more shifts and changes between the gendered job markets is somewhat blurred and partly unexamined. The growth of the entrepreneurial professions (Reed 1996; Poutanen and Kovalainen 2016; Young and Muller 2014), refers to the increasing number of independent expert jobs and occupations, such as business consultants, computer and IT analysts and programmers and engineers, and their increasing role in the new economy. The potential and possibility that lies in the digital services and platforms are undoubtedly changing the ways labour markets function, and this change takes place globally. The rise of the professional services on freelance basis and the new demands for short-term services by households and businesses alike are likely to increase the forms of 'gig economy' as described above. While the professions change and transform, and new professions and jobs arise with the new economy, their gendered arrangements do not necessarily change so much. The gendered formation of new economy and its innovative professions will be discussed in the next chapter.

REFERENCES

Acemoglu, D., & Autor, D. (2011) Skills, tasks and technologies: implications for employment and earnings. In O. Aschenfelter & D. Card (eds.) *Handbook of Labor Economics*. North Holland: Elsevier. 1043–1171.

Acker, J. (2004) Gender, capitalism and globalization. *Critical Sociology*, 30(1): 17–41.

Acs, Z. J., & Audretsch, D. B. (2005) *Entrepreneurship, Innovation and Technological Change*. Boston, USA, Delft, Europe: NOW Publishers ltd.

Adams, Z., & Deakin, S. (2014) Institutional solutions to precariousness and inequality in labour markets. *British Journal of Industrial Relations*, 52(4): 779–809.

Agnew, J. (2001) The new global economy: time-space compression, geopolitics, and global uneven development. *Journal of World-Systems Research*, VII(2): 133–156.

Arthur, W. B. (2007) The structure of invention. *Research Policy*, 36(2007): 274–287.

Atkinson, A. (2004) Top Incomes over the Twentieth Century. LSE public lecture, Speech delivered 20th January, 2004. Old Theatre, LSE.

Atkinson, R. D. (2012) *The Innovation Economics: The Race for Global Advantage*. Yale: Yale University Press.

Atkinson, R. D., & Andes, S. (2010) *The 2010 State New Economy Index*. The Information Technology & Innovation Foundation. Kauffman foundation. Washington: ITIF.

Atkinson, R. D., & Nager, A. B. (2014) *The 2014 State New Economy Index. The Information Technology & Innovation Foundation*. Washington: ITIF.

Autio, E., Kenney, M., Mustar, P., Siegel, D., & Wright, M. (2014) Entrepreneurial innovation: the importance of context. *Research Policy*, 43(7): 1097–1108.

Baber, J. (2001) Globalization and scientific research: the emerging triple helix of state-industry-university relations in Japan and Singapore. *Bulletin of Science Technology & Society*, 21: 401–408.

Bair, J. (2010) On difference and capital: gender and the globalization of production. *Signs*, 36(1): 203–226.

Baldry, C., Bain, P., Taylor, P., Hyman, J., Scholarios, D., Marks, A., Watson, A., Gilbert, K., Gall, G., & Dirk, B. (2007) *The Meaning of work in the New Economy*. London: Palgrave Macmillan.

Balsamo, A. (2014) Gendering the technological imagination. In E. Waltraud & I. Horwath (eds.) *Gender in Science and Technology: Interdisciplinary Approaches*. Bielefeld: Transkript verlag.

Beardi, C. (2015) Eastman Kodak Company. In J. McDonough & K. Egolf (eds.) *The Advertising Age Encyclopedia of Advertising*. London: Routledge.

Beck, U. (2000) *The Brave New World of Work*. Cambridge: Polity.

Beede, D., Julian, T., Langdon, D., McKittrick, G., Khan, B., & Doms, M. (2011) *Women in STEM: A Gender Gap to Innovation*. U.S. Department of Commerce, Economics and Statistics Administration. Washington: ESA.

Van Beers, C., Kleinknecht, A., Ortt, R., & Vertburg, R. (2008) Introduction. In C. Van Beers, A. Kleinknecht, R. Ortt, & R. Vertburg (eds.) *Determinants of Innovative Behaviour. A Firm's Internal Practices and its External Environment*. Basingstoke: Palgrave. 1–12.

Benioff, M. R., & Alder, C. (2009) *Behind the Cloud*. San Francisco: Jossey-Bass.

Berkun, S. (2010) *The Myth of Innovation*. Sebastopol, CA: O´Reilly Media.

Berner, B. (2008) Working knowledge as performance: on the practical understanding of machines. *Work, employment and society*, 22(2): 319–336.

Bernhardt, A. (2014) Labor standards and the reorganization of the work: gaps in data and in research. ILRE Working paper No. 100-14. http://irle.berkeley.edu/workingpapers/100-14.pdf. Retrieved 1.9.2016.

Blake, M. K., & Hanson, S. (2005) Rethinking Innovation: context and gender. *Environment and Planning A*, 37: 681–701.

Boltanski, L., & Chiapello, E. (2006) *The New Spirit of Capitalism*. London: Verso.

Bourdieu, P. (1984) *Distinction: A Social Critique of the Judgement of Taste*. London: Routledge.

Braunerhjelm, P. (2012) Innovation and growth: a technical or entrepreneurial residual. In M. Andersson, B. Johansson, C. Carlsson, & H. Lööf (eds.) *Innovation and Growth: From R&D Strategies of Innovating Firms to Economy-wide Technological Change*. Oxford: Oxford University Press. 286–318.

Breznitz, D., & Zysman, J. (2013) Introduction. In D. Breznitz & J. Zysman (eds.) *The Third Globalization: Can Wealthy Nations Stay Rich in the Twenty-First Century?*. Oxford, UK, New York, US: Oxford University Press.

Brinkley, I. (2016) In search of the Gig Economy. The Work Foundation, August 2016. www.workfoundation.com. Retrieved 28.9.2016.

Brush, C. (2014) *Diana Report 2014*. Babson: Babson College.

Brush, C., Carter, N. M., Gatewood, E. J., Greene, P. G., & Hart, M. (2009) The Diana Project: women business owners and equity capital: the myths dispelled. Babson College Center for Entrepreneurship Research Paper No. 2009–11. Babson College.

Brynjolfsson, E., & McAfee, A. (2014) *The Second Machine Age: Work, Progress, and Prosperity in a Time of Brilliant Technologies*. New York: W. W. Norton.

Brynjolfsson, E., McAfee, A., & Spence, M. (2014) *Labor, Capital and Ideas in the Power Law Economy*. Foreign Affairs. July/August 2014.

Business Insider (2014) Instagram's Kevin Systrom: People keep Asking if my $1 Billion was too small. (www.businessinsider.com Jul. 19, 2014). Retrieved 1.6.2015.

Caraway, T. L. (2007) *Assembing Women: The Feminization of Global Manufacturing*. Ithaca, NY: Cornell University Press.

Carayannis, E. G., Dubina, I. N., & Ilinova, A. A. (2015) Licencing in the context of entrepreneurial university: an empirical evidence and theoretical model. *Journal of Knowledge Economy*, 6: 1–12.

Carr, N. (2009) *The Big Switch: Rewiring the World, from Edison to Google*. New York: W.W. Norton & Company.

Carrigan, C., Quinn, K., & Riskin, E. A. (2011) The gendered division of labor among STEM faculty and the effects of critical mass. *Journal of Diversity in Higher Education*, 4(3): 131–146.

Castells, M. (1996) *The Rise of the Network Society*. Oxford: Blackwell.

Centeno, M. A., & Cohen, J. N. (2010) *Global Capitalism: A Sociological Perspective*. Cambridge, Malden, MA: Polity Press.

Clark, P. F., Stewart, J. B., & Clark, D. A. (2006) The globalization of the labour market for health-care professionals. *International Labor Review*, 145(3): 37–64.

Clarysse, B., Wright, M., Lockett, A., Mustar, P., & Knockaert, M. (2007) Academic spin-offs, formal technology transfer and capital raising. *Industrial and Corporate Change*, 16(4): 609–640.

Close, S. (2016) The political economy of creative entrepreneurship on digital platforms: case study of Etsy.com.Paper presentation at the 49th Hawaii International Conference on System Sciences, 2016. https://www.compu ter.org/csdl/proceedings/hicss/2016/5670/00/5670b901.pdf. Retrieved 11.12.2016.

Cooke, P. (2002) *Knowledge Economies*. New York, London: Routledge.

Cooper, M. (2003) The KodakGirl Collection. www.kodakgirl.com. Retrieved 14.7.2016.

Coyle, D. (2011) *The Economics of Enough: How to Run the Economy As If The Future Matters*. Princeton and Oxford: Princeton University Press.

Daguerre, A. (2014) New corporate elites and the erosion of the Keynesian social compact. *Work, Employment and Society*, 28(2): 323–334.

Diaz Garcia, M.-C., & Welter, F. (2013) Gender identities and practices: inter-preting women entrepreneurs' narratives. *International Small Business Journal*, 31(4): 384–404.

Ding, W. W., Murray, F., & Stuart, T. E. (2006) Gender differences in patenting in the academic life sciences. *Science*, 313: 665–667.

Dolgin, A. (2012) *Manifesto of the New Economy*. Heidelberg: Springer.

Dourish, P., & Bell, G. (2011) *Divining a Digital Future*. Cambridge, MA: MIT Press.

Edgell, R. A., & Vogl, R. (2013) A theory of innovation: benefit, harm, and legal regimes. *Law, Innovation and Technology*, 5(1): 21–53.

Elzinga, A. (2004) Metaphors, models and reification in science and technology policy discourse. *Science as Culture*, 13(1): 105–121.

Etzkowitz, H., & Leydesdorff, L. (2000) The dynamics of innovation: from national systems and 'Mode 2' to a triple helix of University-Industry-Government relations. *Research Policy*, 29(2): 109–123.

Eurofound (2016) Annual Work Programme 2016. European Foundation for the Improvement of Living and Working Conditions. http://www.eurofound.europa.eu/publications/work-programme/2015/annual-work-programme-2016. Retrieved 11.8.2016.

Evans, D. S., & Schmalensee, R. (2016) *Matchmakers: The New Economics of Platform Business*. Boston, MA: Harvard Business Review Press.

EY (2014a) *Adapting and Evolving: Global Venture Capital Insights and Trends*. Brussels: EYGM Ltd.

EY (2014b) *Creating Growth. Measuring Cultural and Creative Markets in the EU*. Brussels: EYGM Ltd.

EY (2015) *Cultural Times. The First Global Map of Cultural and Creative Industries*. Brussels: EYGM Ltd.

Feldman, M. S. (2000) Organizational routines as a source of continuous change. *Organization Science*, 11(6): 611–629.

Felstiner, A. (2011) Working the crowd: employment and labor law in the crowdsourcing industry. *Berkeley Journal of Employment and Labor Law*, 32(1): 143–203.

Flew, T. (2007) *New Media: An Introduction*. Oxford: Oxford University Press.

Florida, R. (2014) Europe in the creative age, revisited. *Demos Quarterly*, 1, 2014–2014. Winter. http://quarterly.demos.co.uk/article/issue-1/europe-in-the-creative-age-revisited-7/. Retrieved 11.12.2016.

Frey, C. B., & Osborne, M. A. (2013) The Future of Employment: How Susceptible are Jobs to Computerisation? Oxford Martin Programme on Technology and Employment. Oxford Martin School, University of Oxford. http://www.futuretech.ox.ac.uk/sites/futuretech.ox.ac.uk/files/The_Future_of_Employment_OMS_Working_Paper_0.pdf.

Gass, R. (2008) Innovation and globalisation: OECD through its looking glass. *OECD Observer* 270/271 December.

Geels, F. W. (2014) Reconceptualizing the co-evolution of firms-in-industries and their environments: developing an inter-disciplinary Triple Embeddedness Framework. *Research Policy*, 43: 261–277.

Goos, M., Manning, A., & Salomons, A. (2014) Explaining job polarization: routine-biased technological change and offshoring. *American Economic Review*, 104(8): 2509–2526.

Gordon, R. (2016) *The Rise and Fall of American Growth: The U.S. Standard of Living since the Civil War*. Princeton: Princeton University Press.

Granovetter, M. (1985) Economic action and social structure: the problem of embeddedness. *American Journal of Sociology*, 91: 481–510.

Greene, P. G., Brush, C. G., Hart, M. M., & Saparito, P. (2001) Patterns of venture capital funding: is gender a factor?. *Venture Capital*, 3(1): 63–83.

Hardin, C. (2014) Finding the 'Neo' in Neoliberalism. *Cultural Studies*, 28(2): 199–221.

Harvey, D. (2005) *A Brief History of Neoliberalism*. Oxford, New York: Oxford University Press.

Hathaway, I. (2015) The Gig Economy is real if you know where to look. Harward Business Review. Aug. 15, 2015.

Hathaway, I., & Muro, M. (2016) Tracking the gig economy: new numbers. Research report. October 13, 2016. Brookings Institute.

Hewlett, S. A. (2007) *Off-Ramps and On-Ramps: Keeping Talented Women on the Road to Success*. Cambridge: Harvard Business School Press.

Hirschfield, L. E. (2015) "I just did everything physically possible to get in there". How men and women chemists enact masculinity differently. *Social Currents*, 2(4): 324–340.

Hong, L., & Page, S. E. (2004) Groups of diverse problem solvers can outperform groups of high-ability problem solvers. *Papers of the National Academy of Sciences, PNAS*, 101(46): 16385–16389.

Howard, A. (2016) *Why fixing tech's gender and racial gaps is more crucial than ever*. Techrepublic.com/. http://www.techrepublic.com/article/why-fixing-techs-gender-and-racial-gaps-is-more-crucial-than-ever/. Retrieved 26.7.2016.

Jacobs, J. (1961) *The Death and Life of Great American Cities*. New York: Random House.

Jardins, J. D. (2010) *The Madame Curie Complex: The Hidden History of Women in Science*. New York: Feminist Press.

Kalleberg, A. (2011) *Good Jobs, Bad Jobs: The Rise of Polarized and Precarious Employment Systems in the United States, 1970s–2000s*. New York: Russell Sage Foundation.

Kanter, R. M. (2000) When a thousand flowers bloom: structural, collective and social conditions for innovation in organization. In R. Swedberg (ed.) *Entrepreneurship*. Oxford: Oxford University Press. 167–210.

Katz, L. F., & Krueger, A. B. (2016) The Rise and Nature of Alternative Work Arrangements in the United States, 1995-2015. Working paper. Princeton University.

Kenney, M., & Zysman, J. (2015) Choosing a future in the platform economy: The implications and consequences of digital platforms. Kauffman Foundation New Entrepreneurial Growth Conference, Discussion Paper. Amelia Island Florida – June 18/19, 2015. http://www.brie.berkeley.edu/wp-content/uploads/2015/02/PlatformEconomy2DistributeJune21.pdf Retrieved 21.6.2016.

Kittur, A., Nickerson, J. V., Bernstein, M. S., Gerber, E. M., Shaw, A., Zimmerman, J., Lease, M., & Horton, J. J. (2013) *The Future of Crowd Work*. CSCW '13

Proceedings of the 2013 conference on Computer supported cooperative work. 1301–1318.

Kotiranta, A., Kovalainen, A., & Rouvinen, P. (2010) Female leadership and company profitability. In C. G. Brush, A. De Bruin, E. J. Gatewood, & C. Henry (eds.) *Women Entrepreneurs and the Global Environment for Growth.* Cheltenham: Edward Elgar. 57–72.

Laestadius, S., & Rickne, A. (2012) The theoretical foundation for Swedish innovation policy. In A. Rickne, S. Læstadius, & H. Henry Etzkowitz (eds.) *Innovation Governance in an Open Economy.* Oxon, Abington: Routledge. 18–50.

Lee, N., & Rodiguez-Pose, A. (2014) Creativity, cities, and innovation. *Environment and Planning A,* 36(5): 1139–1159.

Leonard, E. B. (2002) *Women, Technology and Myth of the Progress.* New York: Pearson.

Lewis, J. A. (2008) *Governments and Global Supply Chains: Measuring Performance in a Networked World.* Washington: CSIS.

Li, E. X. (2012) Globalization 2.0. *New Perspectives Quarterly,* 9(1): 40–44.

Lin, C.-P. (2015) Predicating team performance in technology industry: theoretical aspects of social identity and self-regulation. *Technological Forecasting and Social Change,* 98: 13–23.

Link, A. N., Siegel, D. S., & Bozeman, B. (2007) An empirical analysis of the propensity of academics to engage in informal university technology transfer. *Industrial and Corporate Change,* 16(4): 641–655.

Lins, E., & Lutz, E. (2016) Bridging the gender funding gap: do female entrepreneurs have equal access to venture capital?. *International Journal of Entrepreneurship and Small Business,* 27(2/3): 347–365.

Lippman, S. (2008) Rethinking risk in the new economy: age and cohort effects on unemployment and re-employment. *Human Relations,* 61(9): 1259–1292.

Mandl, K. D., Mandel, J. C., & Kohane, I. C. (2015) Driving innovation in health systems through an apps-based information economy. *Cell Systems,* 1(1): 8–13.

Manning, S. (2013) New Silicon Valleys or a new species? Commoditization of knowledge work and the rise of knowledge services clusters. *Research Policy,* 42(2): 379–390.

McDowell, L. (1997) *Capital Culture: Gender at Work in the City.* London: Blackwell.

McDowell, L. (2008a) Thinking through work: Complex inequalities, constructions of difference and trans-national migrants. *Progress in Human Geography,* 32(4): 491–507.

Meng, Y. (2016) Collaboration patterns and patenting: exploring gender distinctions. *Research Policy,* 45: 56–67.

Miles, R. E., & Snow, C. C. (2003) Organization theory and supply chain management: an evolving research perspective. *Journal of Operations Management,* 25(2): 459–463.

Moghadam, V. M. (2000) Economic restructuring and the gender contract: a case study of Jordan. In M. H. Marchard & A. Runyan Sisson (eds.) *Gender and Global restructuring. Sightings, Sites and Resistances.* New York: Routledge.

Moretti, E. (2012) *The New Geography of Jobs.* New York: Houghton Mifflin Harcourt Publishing Company.

Mui, C., & Carroll, P. B. (2013) *The New Killer Apps: How Large Companies Can Out-Innovate Start-Ups.* New York: Cornerloft Press.

Munir, K. (2012) The demise of Kodak: five reasons. *Wall Street Journal*, Feb. 26, 2012. Blogs.wsj.com/sources/2012/02/26/the-demise-of-kodak-five-rea sons. Retrieved 22.11.2015.

Murray, F., Aghion, P., Detriwapont, M., Kovel, J., & Stern, S. (2009) Of mice and academics: examining the effect of openness on innovation. NBER working paper No. 14819. *American Economic Journal: Economic Policy*, 8(1): 212–252.

Murray, F., & Stern, S. (2014) Do formal intellectual property rights hinder the free flow of scientific knowledge? An empirical test of the anti-commons hypothesis. *Journal of Economic Behavior & Organization*, 63(4): 648–687.

Nadeem, S. (2011) *Dead Ringers. How Outsourcing Is Changing the Way Indians Understand Themselves.* Princeton, NJ: Princeton University Press.

NCWGE (2012) National Coalition for Women and Girls in Education. *Title IX at 40: Working to Ensure Gender Equity in Education.* Washington, DC: NCWGE.NCWIT (2016) Statistics on IT workforce. National Center for Women & Information Technology. https://www.ncwit.org/ncwit-fact-sheet. Retrieved 15.7.2016.

Nelson, L. (1993) Epistemological communities. In L. Alcoff & E. Potter (eds.) *Feminist Epistemologies.* New York: Routledge.

OECD (2004) *Science and technology statistics portal.* Retrieved on 15th June 2015 from https://stats.oecd.org/glossary/detail.asp?ID=6267.

OECD (2005) *Oslo Manual: Guidelines for Collecting and Interpreting Innovation Data.* 3rd ed. Paris: OECD.

Oesch, D., & Rodriguez Menes, J. (2011) Upgrading or polarization? Occupational change in Britain, German, Spain and Switzerland, 1990–2008. *Socio-Economic Review*, 9(3): 503–532.

Oldenziel, R. (2004) *Making Technology Masculine: Men, Women and Modern Machines in America 1870–1945.* Amsterdam: Amsterdam University Press.

Orlikowski, W. J. (2008) Using technology and constituting structures: a practice lens for studying technology in organization. In M. S. Ackerman, C. A. Halverson, T. Erickson, & W. A. Kellogg (eds.) *Resources, Co-Evolution and Artifacts. Theory in CSCW.* London: Springer Verlag Ltd. 255–306.

Pajarinen, M., & Rouvinen, P. (2014) Computerization Threatens One Third of Finnish Employment. ETLA Brief. Helsinki: ETLA, The Research Institute of the Finnish Economy.

Perrons, D., Fagan, C., McDowell, L., Ray, K., & Ward, K. (2007) *Gender Divisions and Working Time in the New Economy*. Cheltenham: Edward Elgar Publishing.

Piketty, T. (2014) *Capital in the 21st Century*. Harvard: Harvard University Press.

Poutanen, S., & Kovalainen, A. (2010) Critical theory. In A. Mills, G. Durepos, & E. Wiebe (eds.) *Encyclopedia of Case Study Research*. Thousand Oaks, London, New Delhi: Sage Publications. 261–265.

Poutanen, S., & Kovalainen, A. (2013) Gendering innovation process in an industrial plant – revisiting tokenism, gender and innovation. *International Journal of Gender and Entrepreneurship*, 5(3): 257–274.

Poutanen, S., & Kovalainen, A. (2016) Professionalism and entrepreneurialism. In M. Dent, I. Lynn Bourgeault, J.-L. Denis, & E. Kuhlmann (eds.) *The Routledge Companion to the Professions and Professionalism*. London and New York: Routledge.

Powell, W. W., & Giannella, E. (2010) Collective invention and inventor networks. In B. H. Hall & N. Rosenberg (eds.) *Handbook of the Economics of Innovation*. Vol. 1. North Holland: Elsevier. 575–605.

Power, M. (1997) *The Audit Society. Rituals of Verification*. Oxford: Oxford University Press.

Ramirez, F. O., & Kwak, N. (2015) Women's enrollments in STEM in higher education: cross-national trends, 1970–2010. In W. Pearson Jr., L. C. Frehill, & C. McNeely (eds.) *Advancing Women in Science*. New York: Springer International.

Reed, M. I. (1996) Expert power and control in late modernity: an empirical review and theoretical synthesis. *Organization Studies*, 17(4): 573–397.

Rickne, A., Laestadius, S., & Etzkowitz, H. (eds.) (2012) *Innovation Governance in an Open Economy – Shaping Regional Nodes in a Globalized World*. Routledge: London.

Rider, S., Hasselberg, Y., & Waluszevski, A. (2013) Introduction. In S. Rider, Y. Hasselberg, & A. Waluszevski (eds.) *Transformations in Research, Higher Education and the Academic Marke. The breakdown of scientific thought*. Dortrecht, Heidelberg, New York, London: Springer.

Ridgeway, C. (2011) *Framed by Gender. How Gender Inequality Persists in the Modern World*. Oxford, UK: Oxford University Press.

Ridgeway, C., & Correll, S. J. (2004) Unpacking the gender system: a theoretical perspective on gender beliefs and social relations. *Gender and Society*, 18(4): 510–531.

Rip, A., & Van Der Meulen, B. J. R. (1996) The post-modern research system. *Science and Public Policy*, 23(6): 343–352.

Robins, K. (1991) Tradition and translation: national culture in its global context. In J. Corner & S. Harvey (eds.) *Enterprise and Heritage*. London: Routledge.

Rogers, D. L. (2016) *The Digital Transformation Playbook: Rethink Your Business for the Digital Age*. New York: Columbia University Press.

Rosser, S. V. (2009) The gender gap in patenting. Is technology transfer a feminist issue?. *NWSA Journal*, 21(2): 65–84.

Rouvinen, P., & Kenney, M. (2015) Tervetuloa uusi työ. *Talouselämä*, 33(2): 35–37.

Rubery, J. (2015) *Re-regulating for inclusive labour markets*. Inclusive Labour Markets, Labour Relations and Working Conditions Branch. Conditions of Work and Employment Series No. 65. International Labour Organization. Geneva: ILO.

Salaman, G. (1974) *Community and Occupation: An Exploration of Work/Leisure Relationships*. Cambridge: Cambridge University Press.

Sandberg, S. (2013) *Lean In: Women, Work and the Will to lead*. New York: Alfred A. Knopf.

Sassen, S. (1996) Cities and communities in the global economy. *American Behavioral Scientists*, 39(5): 629–639.

Sastre, F. (2016) Gender diversity and knowledge innovation barriers. *International Journal of Entrepreneurship and Small Business*, 27(2/3): 193–212.

Saxenian, A. (1994) *Regional Advantage: Culture and Competition in Silicon Valley and Route 128*. Cambridge: Cambridge University Press.

Schiebinger, L. (2008) Getting more women into science and engineering – knowledge issues. In L. Schiebinger (ed.), *Gendered Innovations in Science and Engineering*. Stanford: Stanford University Press. 1–21.

Schmidt, B. (2014) Women, research and universities: excellence without gender bias. In B. Thege, S. Popescu-Willigmann, R. Pioch, & S. Badri-Höher (eds.) *Paths to Career and Success for Women in Science. Findings from International Research*. Wiesbaden: Springer Verlag. 93–116.

Schonfield, E. (2011) The rise of the "creative" class. *Techcrunch. com*. 14.12.2011. at http://techcrunch.com/2011/12/14/creative-class/. Retrieved 1.5.2016.

Sennett, R. (1998) *The Corrosion of Character. The Personal Consequences of Work in the New Capitalism*. New York, London: W.W. Norton & Company.

Sennett, R. (2006) *The Culture of the New Capitalism*. New Haven & London: Yale University Press.

Sennett, R., & Cobb, J. (1972) *The Hidden Injuries of Class*. New York: Knopf, 1972. Re-issued New York: Norton, 1993. Re-issued New Haven: Yale University Press, 2008.

Seppälä, T., & Mattila, J. (2016) Ubiquitous network of systems. Working paper. Berkeley Roundtable on the international economy (BRIE) and ETLA, The Research Institute of the Finnish Economy. https://www.etla.fi/wp-content/uploads/BRIE_Sepp%C3%A4l%C3%A4_Mattila-2016.pdf. Retrieved 14.12.2016.

Seppälä, T., Kenney, M., & Ali-Yrkko, J. (2014) Global value chains with transfer pricing: a product-level supply-chain analysis. *Supply Chain Management: an International Journal*, 19: 445–454.

SFGate (2015) *San Francisco Gate*, 20 Aug.

Shaugnessy, H. (2015) *Shift. The User's Guide to the New Economy*. Boise, ID: Tru Publishing.

Shaugnessy, H. (2016) *Platform Disruption Wave. A New Theory of Disruption and the Eclipse of American Power*. Boise, ID: Tru Publishing.

Shehzad, N. (2011) *Dead Ringers: How Outsourcing is Changing the Way Indians Understand Themselves*. Princeton: Princeton University Press.

Siegel, D., & Wright, M. (2014) University technology transfer offices, licensing, and start-ups. In A. Link, D. Siegel, & M. Wright (eds.) *The Chicago Handbook of Academic Entrepreneurship and Technology Transfer*. Chicago: University of Chicago Press.

Simard, C., & Gammal, D. L. (2012) Solutions to recruit technical women. In *Anita Borg Institute Solutions Series, Anita Borg Institute for Women and Technology*. Palo Alto: Anita Borg Institute.

Simmie, J., Sennett, J., Wood, P., & Hart, D. (2002) Innovation in Europe: a tale of networks, knowledge and trade in five cities. *Regional Studies*, 36: 47–64.

Simon, P. (2011a) *The New Small*. Caldwell, NJ: Motion Publishing.

Simon, P. (2011b) *The Age of the Platform. How Amazon, Apple, Facebook and Google Have Redefined Business*. Las Vegas: Motion Publishing.

Solow, R. (1987) We'd better watch out. *New York Times Book Review* (July 12) 36.

Statista (2015)The Statistics Portal. Venture capital - Statistics & Facts. https://www.statista.com/topics/2565/venture-capital/ Retrieved 14.3.2016.

Stephan, P. E., & Levin, S. G. (2001) Exceptional contributions to US science by foreign-born and freign-educated. *Population Research and Policy Review*, 20(1–2): 59–79.

Stephany, A. (2015) *The Business of Sharing: Making it in the New Sharing Economy*. London: Palgrave Macmillan.

Sundararajan, A. (2016) *The Sharing Economy: The End of Employment and the Rise of Crowd-Based Capitalism*. Cambridge, Massachusetts, London, England: The MIT Press.

Sweet, S., & Meiksins, P. (2012) *Changing Contours of Work. Jobs and Opportunities in the New Economy*. 2 ed. Los Angeles, London, New Delhi, Singapore, Washington DC: SAGE Publications.

Taylor, P. (2010) The globalization of service work: analysing the transnational call centre value chain. In P. Thompson & C. Smith (eds.) *Working Life. Renewing Labour Process Analysis*. Hampshire: Palgrave. 244–268.

The Economist (2003) *The new "new economy"*, Sept.11th, 2003. Special report.

The Economist (2012) *Kodak*. www.theeconomist.com. Retrieved 11.5.2016.

Thursfield, D. (2015) Resistance to teamworking in a UK research and development laboratory. *Work Employment & Society*, 29(6): 989–1006.

Tinkler, J. E., Bunker-Whittington, K., Ku, M. C., & Rees-Davies, A. (2015) Gender and venture capital decision-making: The effects of technical background and social capital on entrepreneurial evaluations. *Social Science Research*, 51(2015): 1–16.

Truss, C., Conway, E., d'Amato, A., Kelly, G., Monks, K., Hannon, E., & Flood, P. C. (2012) Knowledge work: gender-blind or gender-biased?. *Work Employment & Society*, 26(5): 735–754.

Vallance, P. (2015) Design employment in UK regional economies: Industrial and occupational approaches. *Local Economy*, 30(6): 650.

Vallas, S. (2011) *Work: A critique*. Boston: Polity Books.

Webster, J. (2013) *Shaping Women's Work: Gender, Employment and Information Technology*. London: Palgrave.

West, N. (2000) *Kodak and the Lens of Nostalgia*. Charlottesville, VA: The University of Virginia Press.

Wilton, N. (2011) Do employability skills really matter in the UK graduate labour market? The case of business and management graduates. *Work, Employment and society*, 25(1): 85–100.

Womens VC Fund (2016) retrieved 12th January 2016 from www.womensvc fund.com.

World Bank (2016) *World Development Report 2016: Digital Dividends*. Washington: International Bank for Reconstruction and Development, World Bank.

World Economic Forum and Boston Consulting Group (2016) Health Systems Leapfrogging in Emerging Economies: Ecosystem of Partnerships for Leapfrogging. May 2016. Retrieved 12. 10. 2016 at: https://www.bcgperspec tives.com/Images/WEF_Health_Systems_Leapfrogging_Emerging_ Economies_report.pdf.

www.couchsurfing.com (2016). Retrieved 2.10.2016.

www.etsy.com (2016). Retrieved 15.11.2016.

Wyer, M., Barbercheck, M., Cookmeyer, D., Örün Öztürk, H., & Wayne, M. (2013) Introduction. In M. Wyer, M. Barbercheck, D. Cookmeyer, H. Örün Öztürk, & M. Wayne (eds.) *Women, Science and Technology: A Reader in Feminist Science Studies*. New York: Routledge.

Yeung, H. W.-C. (2003) *Chinese Capitalism in a Global Era*. London, New York: Routledge.

Young, M., & Muller, J. (2014) From the sociology of professions to the sociology of professional knowledge. In M. Young & J. Muller (eds.) *Knowledge, Expertise and the Professions*. London: Routledge. 148–151.

Zelizer, V. (1987) *Morals and Markets: The Development of Life Insurance in the United States*. New York: Columbia University Press.

Ziman, J. (1994) *Prometheus Bound: Science in a Dynamic Steady State*. Cambridge: Cambridge University Press.

Zumbrun, J., & Sussman, A. L. (2015) Proof of a 'Gig economy' revolution is hard to find. *The Wall Street Journal*. 26. 7. 2015. at https://search.proquest. com/docview/1698834883?accountid=14774. Retrieved 8.5.2016.

CHAPTER 4

Innovations, Gender and the New Economy

Digitalization is one central feature common to various aspects of the new economy, be this business-to-business activity, production networks or consumption in the new economy. Digitalization has come to mean many things at the same time, but perhaps the most prevalent characteristic of the digitalization of the economy is its platform function. Digital platforms enable the rapid spread of new developments and new features of innovation systems irrespective of whether they are new scientific innovations or new commercial innovations, or both. Since the rapid spread of innovative Internet technologies, the collective understanding of human identity has also become a pertinent issue. Gender is one key element in this discussion. When the economy becomes digital, how is gender understood and subsumed in the new economy? How and in what ways does gender relate to and become enacted in the digital new economy?

There are several ways to determine and analyse the gendering of the digital economy, taking into account the complex nature of categories, such as gender, race, class and geography, as these all play out in the context of the digital economy. The digital space allows several forms of gender identities to emerge, and gender is an overriding aspect in all those identities. Gender becomes differently formulated, presented, displayed and analysed in the different contexts of the new economy. Different aspects of this economy, such as games development and gaming, or customer interfaces and user experiences, or the digital-based drug development industry, all exemplify a very different set of gender issues. Yet

© The Author(s) 2017
S. Poutanen, A. Kovalainen, *Gender and Innovation in the New Economy*, DOI 10.1057/978-1-137-52702-8_4

they all have importance in relation to innovations and ways innovations are developed, promoted and adopted. This chapter will provide some openings to the topic.

4.1   GENDER AND GAMING INDUSTRIES

The economic innovations in games development and gaming are different by nature to the innovations described in earlier chapters of this book. Digital and Internet enabled innovations transform the very basis of the economic structure through "permission-less innovation" – that is, innovation by many individuals, groups, networks and start-ups (Thierer 2014). How and in what ways gender and innovations relate to the new platform economies, are the key questions addressed in this chapter. This chapter delves into the nature of innovative platforms in the new economy and the ways in which gender becomes part of the new economic landscape, digital development and global on-line economic space.

The very short history of computer games extends to the first generation of video games which were invented in the 1970s. Before that, individual experiments are to be found, but with no commercial spread. During the 1970s, the first generation of home consoles emerged, creating mass markets for the early video games. During the 1980s, the first gaming computers, early online gaming opportunities and handheld LCD games emerged, to be followed by several computer models such as IBM PC, Atari, Commodore, and Apple Macintosh computers as prime gaming computers during the 1980s. Shareware gaming started and achieved its first big success in the 1990s, as did the first versions of 3D graphics in the games (Egenfeldt-Nielsen et al. 2016). The new and increasingly competitive market for portable game systems came with handheld devices, Nokia being the very first to install a mobile game (Snake) onto a phone in 1997 (Wright 2016). With the development of 3D graphics, video and computer games enthralled their audiences, and gained new consumers.

During the early years of the twenty-first century and with the affordable broadband Internet connectivity spreading in most countries, online gaming became popular and indeed was seen as one major way of innovating in the gaming industry. As one consequence of this, social network gaming has become a hugely popular form of games and gaming, offering new avenues for social interaction and connection (e.g. Lowood 2006; Harvey 2015). These loose ties through gaming enabled for their part the development of other kinds of social networks (e.g. Abbate 2012; Adams and Demaiter 2008).

In relation to games and gaming, the first aspects of feminism, gender research and feminist research all emphasized the growing importance of technological literacy and the growing need for and creation of computer games designed especially for girls (Harvey 2015). This idea of 'girls' games' was especially prevalent in the 1990s research focusing on gender and games, but also later several research projects have confirmed that it does matter who designs the games, and for whom the games are designed. The results of a 3-year longitudinal study conducted at the Michigan State University, with a mixed-method design, showed that, in several ways, the designer's gender influenced the desirability of a game among 5th to 8th graders in schools, irrespective of the type of the game. The gender of the designer seems to have a significant influence on the design outcome of games. Boys overwhelmingly picked games based entirely on fighting as their top ranked games. Girls overwhelmingly ranked those same fighting games as their least preferred ones (Heeter et al. 2009).

The studies on gender and game play in virtual on-line games have so far seldom focused on the perspective of boys. A study which focused on boys' own views and participation in a tween-centric virtual world game showed that virtual worlds offer space for the expression of boys' culture, but that they are qualitatively distinct from other gaming environments and perhaps need to be studied in their own terms (Bogg and Prescott 2014; Searle and Kafai 2012; Delamere and Shaw 2008).

Studies that have emphasized the lucidity and contextual nature of the gaming experiences more generally do not put strict borders between the type of game and gaming. These studies obtained different results in comparison to the Michigan State University study. One study analysing the computer gaming preferences of girls in a games club in an all-girl school in the UK argues that "gaming tastes" are alterable, varying and site specific. Researchers have concluded that gaming preferences may relate to the attributes of particular games, but that they are also highly dependent on the player's recognition and knowledge of these specific attributes, which for their part are shaped by gender. The several interconnections show the complexity of analysing gender. Research has in general evidenced that the gaming preferences may manifest along gendered lines, but contextual and practice-related matters do play a significant part in the process (Royse et al. 2007; Carr 2005).

Consumption studies have further brought forward the complexity of discussing gender and the consumption of games and practices in gaming. In some consumption studies, individual differences in the consumption of computer games have been found to intersect heavily with gender. Indeed, research has evidenced that men play video games more than women (e.g. Winn and Heeter 2009), but women are a growing group of gamers, and within the 'group of women' several different patterns of consumption are found. Indeed, when analysing women's gaming, important features blur the rigid and stereotypical 'gender as difference' results: A study by Royse, Lee, Undrahbyan and Consalvo focused on adult women and their gaming habits and brought forward particular attention to the differences between women in the levels of play, as well as genre preferences for the games played. Researchers identified three different levels in game consumption. For the so-called "power gamers", technology and gender were most highly integrated. Women who played a lot also enjoyed multiple pleasures from the gaming experience itself, including the technical mastery of game-based skills and techniques and also including winning competitions in gaming. Women who were moderate gamers did not focus on technicalities, but played games for several reasons, and seldom due to competition. For most moderate players the main motivation in gaming was to master the game and through that mastery cope with their real lives and hence, they reported taking pleasure in controlling the gaming environment (Royse et al. 2007).

The contextual factors that explain part of the gender disparity in gaming is one explanation for the rarity of women among the players and designers. More recently, new representations of gender, culture, diversity, sexuality, and masculinity have risen in the study of games, gaming practices and gender (e.g. Harvey 2015; Jenson and De Castell 2010; Shaw 2010). Games and gaming are part of a wider media landscape on the one hand, and a subset of a much larger set of social practices on the other. Games and gaming also encompass social communities irrespective of their nature. Reproducing the gendering in games and gaming is thus not an exception or deviation from other social and economic settings. The topic of gender is therefore crucial to address, without essentializing gender differences and homogenizing the category of "woman" (Jenson and De Castell 2014). Research results have for a long time shown that highly gendered roles in gaming are also socially constructed: for example in public settings, such as mixed-gender game clubs, boys dominate if the gaming culture has from the beginning been more supportive to the boys' active role in gaming (e.g. Voorhees 2014; Bryce et al. 2006). Yet, they are far from being settled into specific positions, even if gender differences have been found in the use of time in each game playing session (e.g. Gwee et al. 2013) and the ways game is gendered. This construction seems to hold true even when on-line chess is studied (Basaanjav 2016).

The gendering of digital games and gaming takes many forms. Most prevalent sorts of gendering, such as the gendering of gaming scenes and gaming communities, as well as gender differences in gaming, are often attributed to the gender differences of the game players in the majority of studies. A study on Swedish gaming discussion showed while there was a consensus for welcoming both genders into the gaming community, and the absence of girls was seen as a problem, within the competitive gaming gender was not viewed so positively (Svenigsson 2012). More action research oriented studies aim to develop games with an educational and transformative edge.

The enormous educational potential of the digital gaming and games industry has been increasingly discussed since the 1990s, but the gendered nature of the games developed by the gaming industry has remained a matter of controversy (e.g. Prescott and McGurren 2014; Sigurdadottir et al. 2015). The PISA study examines the scholastic performance of 15-year-old school pupils' in mathematics, science and reading every third year (OECD 2014). The results from the global PISA studies show in general that moderate use of single-player games is associated with an individual performance advantage. Several research reports show that there are gender differences between boys and girls in gaming which are prevalent already in school and in the use of educational games and gaming (e.g. Borgonovi 2016; Joiner et al. 2011).

In spite of the growth and spread of game studies overall in the last decade, the research into computer games is still in its infancy in the academic world when it comes to the economic side of gaming. Many of the robust longitudinal statistics on games, their players and the intensity of the gaming are simply missing. Also, much of the data compiled is time limited or national by nature and lacking most aspects for comparative research. For that reason, it is difficult to create time spanning cross-national comparisons. On the other hand, the national boundaries may not have crucial importance in the emergence of clusters and local and transnational ecosystems of digital game players' communities.

The focus on the attributes of players has shifted from individuals to the process of gaming, the complex interplay between the gamer and the game, and the group dynamism, for example. Dovey and Kennedy (2006) and Israel et al. (2016) argue that in gaming it is not necessarily gender, race or age that would determine the inclusion of an individual or a group, but what they call "technicity", mastery of the game and a network of technological competences that often are associated with white males and masculine culture, and exemplified by hackers and so-called "hardcore" gamers. As reported in many studies (e.g. Lin 2008; Downey 2012), the virtual spaces of games are still often marked as distinctively masculine, thus reinforcing the masculinity of gaming and the games themselves. The gender divisions, and with them the gendering of the games, are shifting: current research shows that the gender distribution of game players has evened out and the share of adult players has grown remarkably (Pierce 2015). The mastery of the technicity is not gender-dependent, but more, time- and intensity-bound: the more time used for playing, the better mastery of the technicity can be achieved. It is then a cultural question as to whether culturally equal amounts of time are allowed for boys and for girls to play games and immerse themselves in game playing.

Two surveys analysed the demographics of gamers in 2010 and found that 55% of the social network gaming demographic in the USA consisted of women, while in UK women made up nearly 60% of social network gaming demographic (Ingram 2014). In addition, most social gamers in the 2010 study were around the 30–59 age range, with the average social gamer being 43 years old. Based on these and other results it has been suggested that the reason why social games may appeal more to the older demographic is because the social games are free, and easier to grasp and advance through in a short period of time, and do not involve as much violence as traditional video games. Shaw on the other hand argues in her study of gamers that game players experience race, gender, and sexuality concurrently. She shows how representations and design matter to gamers (Shaw 2015). Being a gamer can result from having higher skills in gaming and for establishing connections with other gamers and receiving respect in the communities of gamers, the skills matter greatly.

The widening field of game playing, as well as the adoption of learning games as part of the education curriculum – and more broadly – the gamification of education (e.g. Ochsner et al. 2015), as well shifts in the game players' demographics all mean that the patterns of games consumption are constantly shifting and broadening as well. These shifts and changes undoubtedly increase the need for variety in game planning and design, and call for more gender aware game designs and applications for games and learning. Contemporary popular culture is constantly feeding into games development and vice versa. Games have high degree of influence on popular culture in many ways. Games and game series, such as: "Super Mario Bros.," "Mortal Combat," "Hitman: Agent 47," "Lara Croft: Tomb Raider," "Max Payne," and "Prince of Persia: The Sands of Time," were all made into movies. In addition, several films have their plots centred on video games.

In general, videogames have a strong influence on popular culture in many innovative ways. Interactive culture – shown through and with the help of the platform economy – is essentially also lowering the thresholds between the different forms of culture. Games and gaming have changed the ways many other forms of culture, such as film and music are produced and consumed, and indeed, the boundaries between the different media forms are blurring. Games create symbolic worlds and subcultures, and with them identities and group affiliations through their consumption (Dovey and Kennedy 2006; Shaw 2010).

One example of the ways it is possible to spread the use of games and gaming in education is research initiated and funded game development. Educational games are designed to assist the learning of specific subjects or skills through game play. Game-based learning has defined learning outcomes and it inevitably changes the ways learning takes place. Games can bring together ways of knowing and doing, and some forms of games provide an adaptive web-based exercise system that generates problems or questions for students based on their skills and performance. Some of educational games have recently become commercially successful.

Research based games are usually innovative games which target those issues and knowledge gaps, and even biases that often are unrecognized and not dealt with in gaming. Of such games, "Fair Play" (www.gameslearningsociety.org) – is one example, which was developed in research in the NIH Pathfinder – grant funded project at the Wisconsin Institute for Discovery, at the University of Wisconsin-Madison. The game is based on the research proven impact of game-based learning on attitudes, behaviours and social interactions. The game is one example of the collaborative innovations of the public and private sectors. The public sector financed the project and the universities developed the game as part of their research. This has resulted in academic publications and has had an awareness raising impact in academia which is helping to change the culture surrounding games and gaming (Gutierrez et al. 2014).

A 20-year history of research on gender and games originally demonstrated that women and girls did not often have the same access to or interest in digital games and game culture as men and boys did. Furthermore, the research has established that women are underrepresented and stereotyped, often hyper-sexualized within the game content (Downs and Smith 2010; Kidd and Turner 2016). The masculinities of the gaming culture were and still are, to a large extent hegemonic, leaving little space for women (Kidd and Turner 2016). However, more recent research shows that even these gender differences in abilities and perceived abilities are often diminished once females have had the opportunity to train and engage in gaming (e.g. Jenson and de Castell 2010). Furthermore, research has shown that it is not only the games themselves and their portrayal of gender, but also the ways games and games marketing portray the social context or show lack of diversity, and this representation can have negative effects on efficacy for marginalized gamers (e.g. Williams et al. 2009).

Since 2000s and up until now, the number of female players has globally grown and the number of women as games designers has increased from the early days of digital games. This increase has not been visible, in general, in the number of girls and young women completing a degree in engineering or technology programs. A report from the National Center for Women & Information Technology in the USA shows that slightly over half of all US women who enter technology fields leave their employers in their mid-careers (Ashcraft and Blithe 2010). Some researchers point out that the reasons for leaving the labour force in the USA are for child-bearing and family reasons to the extent that children explain much of the differences in remaining or leaving technical careers in most age cohorts (Kahn and Ginther 2015; Ginther and Kahn 2015).

One of the key elements in current society is the rapid spread of technology in all facets of life and in various fields and forms. The spread of technology is no longer based on hardware alone, but increasingly in the multitudes of apps, their networks, and networked devices and infrastructure. The consumer acceptance and adoption of apps is highly crucial for all apps. Following this, the increasing attention being paid to identify the politics of the users and consumers, in addition to gender politics in general and pressure groups, have all led to demands for changes in the digital world more generally. One example of such changes is the gender categories available on the Internet identity pages. Several of the virtual and digital service providers have taken the gender aspect seriously: Facebook announced in 2014 that instead of identifying as male or female (or hiding gender entirely), US users of Facebook will be able to choose a custom gender, along with a gender-sensitive pronoun that allows for use of the singular "they". Later, Facebook has broadened the selection of gender options, and the users can also write in their own solutions if needed (Gigaom 2014, 2015).

Exploring the innovations in digitalization is complex: is the leading edge of the innovation technological, or is it led by economic innovation, or a combination of the two? Does the design or creative innovation form the leading edge, or is the innovation in the way all of the mentioned elements are combined? Digitalization in games includes and combines art, business, and technology. The potential of the innovative new amalgams of these three fields lies in their critical engagement. This can be found in recent gaming development — the contextualization and the opportunities this provides offer ways to combine the three areas in highly innovative ways.

Augmented reality (AR), and more specifically, mobile augmented reality (MAR), is one example of such contextualization, where the game itself and its characters are global, but the gaming is online and location based and materializes in local settings. Augmented reality combines real and computer-generated digital information with the user's view of the real physical world in such a way that the two appear as one environment. AR is not intended to replace the real world, rather it assumed to supplement it. In research, augmented reality is related to a broader concept of "mixed reality" (MR). Mixed reality refers to the integration and merging of the real and virtual worlds where physical and virtual objects complement and interact with each other (Olsson et al. 2013). Mobile, hand-held devices (mobile phones, digital cameras and navigators) offer a fruitful platform to apply mixed and augmented reality technologies. Indeed, as several researchers have pointed out, the gap between the real world overall and its digital counterparts or related metadata is becoming narrower (Olsson et al. 2013; Hsu et al. 2017).

Augmented reality games, such as globally hugely popular "Pokémon Go" or its predecessor "Ingress" both represent enduring and innovative elements of the new economy. The enduring elements consist of the use of the robust existing technologies and cultural elements (such as the original Pokémon creatures and existing algorithms) and the new innovative elements (such as the use of advanced and new technologies, smartphone cameras, GPS technology and digital maps). The first "real world gaming" platform, such as "Pokémon Go", projects the original Pokémon creatures into the real, local world and surroundings with the help of AR and GPS.

In so doing, it combines mobile location technology and augmented reality in order to create an authentic gaming experience that moves people around in their nearby surroundings. The ways people explore their surroundings by moving around physically in the real world in these augmented reality games benefit their gaming.

The original "Pokémon" game, which was developed in 1996 by Satoshi Tajiri, is based on his childhood hobbies. The game spread quickly into popular and consumer culture, including clothes, comics, toys and movies with emotive ties between players and objects. In that spread, the Pokémon creatures established their place in the new economy and at the same time exemplify the ways in which value is added and created in the new economy in new ways. Value no longer resides in the physical production alone, but in the assemblage of the existing cultural and technological capital. These assemblages take place in new ways, tying the knot between the highly consumerist culture and mobile gaming that adjusts to culture by capitalizing on the additive elements of gaming and the familiarity of the context. An example of the innovative ways to create more sales in shops and cafes for example is through the Pokémon Go players. This means receiving acceptance by Niantic Labs, the Pokémon game developer, to become the pit-stop for Pokémon creatures (Evangelho 2016).

Earlier, AR had been used in glasses and in visual aids, but not in games. Apart from AR games, augmented reality or mixed reality is now being used increasingly in education and in teaching, as the experiment by Hsu et al. (2017) shows. In an experiment by Hsu et al., AR was embedded in the authentic inquiry activities of students so that they could experience when they needed to carry out certain medical surgical procedures, and how to proceed with the actual surgery. More widely, AR is most used in education for science, humanities and arts fields (Bacca et al. 2014), and indeed it could offer innovative solutions to other disciplinary fields.

It is worth exploring how gender emerges in these new settings. The questions of how gender becomes displayed or becomes an active agent are hugely complex questions. In these assemblages of augmented reality, gender relates to the different forms of capital, such as cultural capital and also the market economy. With the innovative augmented reality games, gaming is no longer an isolated hobby to be taken up every now and then, but it may become a life-long commitment with almost endless possibilities for capitalization from the new innovations of the digital companies. Therefore, simply counting the sex of the game players remains an important but alone a poor indicator of the gendered agencies and intersections involved in new augmented reality games, for example. As the gender research of the digital world has moved away from the self-evident categories into a wider view, the associated gendering and its different tenets become the focus of the analysis. If in gaming the playable characters are not understood as merely vehicles for the players but as a set of capabilities or affective attachments and markers for the player's identification (e.g. Burns and Schott 2004; Aarseth 2004), then gender becomes inscribed in the characters and has specific functions. The hegemonic forms of masculinity in game cultures have received a lot of research and media attention (e.g. Burrill 2008), yet the complexity in forms of masculinity has interestingly increased, adding new layers to the ways gender becomes understood in games and gaming (Voorhees 2014).

Digital technologies have been given several gender-loaded tasks, ranging from bodily transcendence to gender politics and to new contexts for identity reconstruction. Computer mediated communication was, from the very beginning, seen to serve "as a place for construction and reconstruction of identity" (Turkle 1995: 14). For some time, digital technologies were also assumed to offer liberation from gender roles and stereotypes (Arvidsson and Foka 2015; Turkle 2011). However, the Internet and digital world do not alone make things happen: the power of the digital is in its ability to create new innovations and to transform and change the material world through those innovations. Furthermore, the renewal of traditional industries through new innovations increases the impact of digitalization. The visible and tangible outcomes in individuals' lives – where computer and Internet presence is very different than, say, ten years ago – are thus "end-products" of a long production value-chain. Through these value-chains companies create surpluses with devices, connections, abilities, apps and networks. The digital transformation thus pervades the entire production chain, from business to business relations and manufacturing contracts down to the consumers. The transformation materializes and is visible in individuals' lives through the consumption of digital services and items.

Looking into the gendering of the gaming and digital platforms in relation to games involves intersectionalities, among them the generation factor (Turkle 2011; Fromme and Unger 2012). Intersecting social identities, biological and social categories interact in multiple and often simultaneous ways. These categories are not 'black boxes' (Poutanen and

Kovalainen 2013; Lutz et al. 2011; Poutanen and Kovalainen 2010) but evolve in relation to each other, in a processual and complex manner (e.g. McCall 2005; Lykke 2011). One concrete example of how the gender becomes transformed into game figures is the so-called 'dadification' of digital games, in contrast to positioning the player as a mother or girl figure Voorhees 2016; Consalvo 2012). The trend where digital game players are positioned either in the role of a father or father figure is growing and widening the options also for the male gender. 'Dadification' is one example of many conscious attempts of directing gaming away from hyper-masculinities in games (Voorhees 2016).

4.2   GENDER AND TECHNICAL DESIGN

There are differences within the technical fields, and gaming is one such field where gender differences are obvious among the games developers and coders. One way to tackle this complex problem is to "normalize" technology as a career for women, and another way is to change the organizational cultures. For both purposes, several NGOs focus on the numbers of female coders and create bottom-up-activities for promoting girls and women into coding and games development (Girlswhocode 2016; Women's coding collective 2016; GraceHopper.org 2016). One way to close the gender gap in technology is to teach girls how to code, and draw them into technical jobs and technical education because tech jobs are among the fastest growing jobs in many countries. The initiatives for girls are plenty as described below.

Non-profit organizations in many countries take care of the additional teaching not covered in the school curriculum. 'Girls Who Code' is such organization which started in New York with 20 girls and from New York spread nationwide to 42 states and with over 10,000 girls participating. Free after-school clubs and summer camps for girls on coding have opened up possibilities for majoring in computer science, and the majority of participants have planned some further studies in relation to computing (girlswhocode.com 2016). The 'Girls Who Code' initiative paired intensive instruction in robotics, web design and mobile development with high level mentorship. The founder and CEO of the 'Girls Who Code', Reshma Saujani, notes that the lack of women in innovative technology and innovation entrepreneurship is not a so-called pipeline problem – women leaving the profession – but a larger question (girlswhocode.com 2016).

Narratives of similar successful alumni stories are told by NGOs and networks in promoting girls into coding. The outcome of their activity is one which has gendered impacts. At the same time both the nature of games and the use of games and gaming has changed, and attention has extended to issues such as the marginalization of professional female players, and the proliferation of stereotypical avatars, for example (Jenkins and Cassell 2008; Jenson and De Castell 2010). Game studies in general have more broadly than earlier examined the socio-cultural dimensions of games and gaming and added complexity and lucidity to gender aspects in games (e.g. Iacovides et al. 2015; Jenson and De Castell 2014; Gray 2013). In a similar vein, the gender studies analysing games and gaming have moved beyond the binary oppositions and rigid positions towards intersectionality, dynamism and practice-oriented approaches in the analyses of digital gender construction.

4.2.1 Case: Getting Women into a Tech-Connected World – the Legacy of Ada Lovelace

A leading idea for sparking enthusiasm and giving tools to girls for building web applications from concept to code lies behind "Hello Ruby." While this idea may open up the web for girls as a platform for their ideas and plans, it also can widen their interests at school towards STEM fields and later network them to a wider community of programming, coders and coding. Coding and programming are not very often major for girls at school, and while coding and programming would need to be part of the curriculum, they seldom are.

One way to tackle the problem is to make coding an interesting and creative thing in itself, with less or little emphasis on mathematics. This is how Linda Liukas came up with her theme for a children's book. The children's book about a girl named Ruby with a big imagination and lots of curiosity towards new things, deals with technology, computers and coding as a form of inspiration for children and was published in 2015. The "Hello Ruby" book that Linda Liukas developed did not remain the only one, as she is already writing a new book. She worked at the Codeacademy in New York as a programmer and did not have the extra time needed to devote to writing. She knew she would need some editorial help and printing assistance to get the book finalized. The crowd-funding site, Kick Starter, helped Ms. Liukas to finalize her children's book about a brave, 6-year-old girl, Ruby, who adventures into the computer world with animal figures and learns how to organize things, that is, to code. The initial target for the Kick Starter funding Liukas set to get her book finalized with some editing help and printing was 10,000 USD. She eventually gathered over 350,000 USD for her book project. This sum initially gave her some freedom and writing time, and helped her to find 10,000 people who were willing to buy her book (Liukas 2016). This in turn helped her to get a publishing contract with MacMillan's children's book unit. In the MacMillan's author space, on their website, Liukas is introduced as a programmer, storyteller and illustrator (www.us.macmil lan.com 2016).

"Hello Ruby" is partly a picture book and partly an activity book which introduces the fundamentals of the computational thinking to children, through innovative and playful ideas, activities and exercises with a target group of 5 to 8-year-old children, not only girls but boys as well. All the creatures in "Hello Ruby" refer to features in the technology world (Liukas 2015). All the characters in the book refer to the tech world: The name of the little girl Ruby originates from programming language Ruby, while messy foxes in the book refer to Mozilla Firefox, penguins refer to Linux, the robots refer to Android by Google and finally, the snow leopards refer to Apple. The aim of the "Hello Ruby" book is not to teach children code writing as such, but more to focus on logical thinking and logics in all activities: in the book Ruby is asked to put her day clothes on, and she follows the order and dresses, but leaves her pyjamas underneath her clothes because her father did not ask her to take her pyjamas off before getting dressed. Ruby learns to think of big problems as a bunch of smaller problems logically connected to each other through problem-solving.

If "Hello Ruby" was an innovative children's book, Rails Girls is another type of innovation by Ms. Liukas. Rails Girls is a social innovation, a non-profit volunteer community that teaches women and girls coding globally. Rails Girls was founded by Liukas and Karri Saarinen in 2010 in Finland, but has since then widened into a global non-profit network to provide tools for building applications and a community of technology with low or no threshold for participation. Ruby Rails is originally a web application development framework, written in the Ruby language. The technical skills needed are minimal as the Ruby language allows the writer to write with little code and is considered particularly suitable for beginners.

Rails Girls bears a resemblance to a programming language, but according to their website, it seems to carry strongly the idea that computational thinking is foundational for everybody. For that purpose, Rails Girls provides guidance for building apps, putting them online, adding profiles, writing games, using several programming languages and in general getting an understanding of what is going on in the web. The Rails Girls workshops, organized by volunteers in over 160 cities, have in a few years taught more than 10,000 women the foundations of programming. In a national news interview Linda Liukas defines herself as the "Bjork of programming who makes cool stuff" with the general aim to make technology more interesting, easily accessible and less dusty and boring (Eskonen 2015).

The innovations briefly described above are very different types of innovations from those innovations that come through scientific inventions and that bring in something new, as described earlier in Chapters 2 and 3. The two innovations by Liukas, first the innovation where learning the coding is developed into a children's book for girls and boys, and the second innovation where she networks volunteers to teach girls and women, are both examples of social innovations rather than traditional business innovations. They are social innovations because both innovations are novel solutions to an existing problem that can be defined as "social". In this case, the "social problem" is the lack of girls and women in programming and in technical occupations and jobs.

But is it 'social' enough, or enough of a 'problem', for this to be considered a true social innovation in this particular way? Many innovations target social problems and needs, but they target them for profit. In most social innovations, the traditional boundaries that separate business innovations from social innovations, and not-for-profits from for-profits, no longer are sustainable. What makes an innovation a social innovation and what distinguishes an innovation from a social innovation? This will be discussed later in the book.

In many interviews, also on her own blog, Liukas and other founders of girls' educational coding training refer to Ada Lovelace as a pioneer in her time in the nineteenth century. Ada is not only the name of an online media journal (journal of gender, new media and technology), but also and more importantly, the name is an integral part of the history of computers and programming. Ada Lovelace (1815–1852) was an English mathematician known for her work on Charles Babbage's mechanical computer. Lovelace's notes and remarks on the engine include what is recognized as the first algorithm intended to be carried out by a machine, a ground-breaking innovation in the form of making the machine, an early mechanical calculator, work in an intelligent manner. As a result of that work, she is often regarded as the first computer programmer in the world, even if not in the modern sense, but in a historical context (Hammerman and Russell 2015). The Lovelace Medal was established by the British Computer Society (BCS) in 1998. The Medal is presented to individuals who have made an outstanding contribution to the understanding or advancement of Computing. Very few women have received the medal over its history (British Computer Society 2016).

Research has evidenced the oblivion of women who were part of the early history of computing and taking the first technological steps in programming, and there is a clear lack of their documented traces. One of the newer arenas to address the issue of missing and forgotten women in the sphere of computer technology are documentary films. One example of these visualizations of women within the field is a documentary film called "The CODE" by Robin Houser Reynolds, which shows the dearth of female and minority software engineers and explores the reasons for the gender gap and digital divide. The film aims to "inspire changes" in mindset, education, and start-up culture, by bringing more women into programming and technical careers. The film shows how this critical gendered digital divide could be closed. The film also revisits the women pioneers of coding and programming from Ada Lovelace to Grace Hopper and the new high-tech gurus such as Susan Wojcicki. Highlighting the remarkable work by the early pioneers is important for the recognition of women in the field. On the website of the film, the director acknowledges as one of the motivating issues for the film the fact that women have been important part of the history of computers but have been almost entirely written out of that history (Code 2016).

The Little Miss Geek campaign in the UK resembles the Rails Girls initiative described in the case above. The Little Miss Geek project aims to inspire girls and young women to take up careers in technology. Little Miss Geek is part of the Lady Geek initiative and campaign in the UK, which has further been turned into a business consulting agency. Lady Geek and Little Miss Geek were started in 2010 by Belinda Parmar, who also wanted to increase women's participation in the technology industry and make technology more accessible to women and girls. The term 'geek' originally referred to person/persons who were very good with technology, and devoted to computers and related technology (Oxford English Dictionary 2016). In addition to meaning someone having good technical skills, the term "geek" often connotes a lack of skills that tend to make people popular among other people, such as fashion or athletics. Nowadays 'geek' refers to a person who is at home and skilled with computers and overall with any kind of technology. The Lady Geek organization is based in London, in the UK. The innovation by Belinda Parmar was to create and introduce an 'empathy' index, used as added value for HRM, PR and recruitment. The empathy index developed by Parmar and her colleagues brings a comparable element to the ways companies work by showing their understanding of others' experiences. The algorithm measures financial data, social media and qualitative data – focusing on customer, employee and social media data, giving each an equal weighting, and as a result, provides a company's 'empathy quotient' (Parmar 2015).

Little Miss Geek is a not-for-profit social enterprise, running tech clubs in inner city state schools (a several-week long program on subjects from robotics to wearable technology) to inspire 13-year-old girls to go on to study computer science. On its website Little Miss Geek announce in their vision that, "We want the next Mark Zuckerberg to be female", (www.theempathybusiness.co.uk 2016).

In a similar manner to Linda Liukas, Belinda Parmar's social innovation also resulted in a book, "Little Miss Geek: Bridging the Gap between Girls and Technology" (Parmar 2012). The book explores the reasons why so many women want to consume technology instead of creating it. Little Miss Geek runs campaigns, such as HERinHero, which is a UK-based campaign on women as trailblazers in technology industries (www.theempathybusiness.co.uk 2016). Little Miss Geek also takes up the subject and utilizes the history of computing, by holding training sessions for UK schools on the Ada Lovelace Day.

The Girls in Tech initiative (GIT), is a global non-profit network focused on raising interest in technology, start-ups, and tech platforms, initiated and started by Ms. Adriana Gascoigne in 2007. GIT aims to boost professional women in high-tech and start-ups, but also in their careers in STEM fields (girlsintech.org 2016). GIT is headquartered in San Francisco but functions globally, relying on volunteers and sponsors. For its activities, GIT relies on volunteer efforts to lead each of the 60 local units. GIT organizes several activities such as the "Catalyst" conference as well as various boot camps, hackathons, women's pitch nights, mentorship programmes and other workshops. GIT also runs an informative website with information about the impact of helping and empowering girls and young women in poor countries (www.ifgirlsrantheworld.com 2016).

Kathryn Parsons, one of the four founders of Decoded, a digital learning company wanted to demystify coding and spread "digital fun". Her start-up firm promised to teach people with zero computing knowledge to build an app in one day, and since its start-up in 2011, it has grown into a global company. Decoded runs one-day courses for company personnel, children, and an education program which instructs teachers on incorporating code into their lessons (Decoded.com 2016).

The four women with innovative solutions in digital learning briefly described above are not only social innovators in relation to girls and ICT/technology. Many initiatives aim for achieving the same solution: adding girls' scholarly and career interests to technology and helping them to develop their present skills in coding, for example. One example of such an initiative is the Hour of code campaign, which is one successful example of a campaign designed for young girls and ICT (e.g. Gilpin 2014).

An example of another type of awareness raising activity is the Girls in ICT day which is an international activity day that has been celebrated several times over the years in over 150 countries. The Girls in ICT initiative is an effort by the International Telecommunication Union (ITU) for raising awareness on empowering and encouraging girls and young women to consider studies and careers in ICT. The organizers of this international activity have developed a portal for girls, with information channels, blogs, and events such as a programming challenge for girls, the "Scientista" symposium and the Africa Summit On Women and Girls in Technology, just to mention a few (www.itu.int 2017).

The Women's Start-up Lab (womenstartuplab.com 2016) follows a rather similar model as all the previous examples: their mission is to empower women to become powerful economic actors, to influence and to shape the world through their companies. Just how that happens, is articulated on the webpage through detailed program takeaways and these activities rely on the participating individuals. The participants are offered workshop instructors, coaches, networking opportunities and pitching, advisors, angel investors and roadmaps. The Women's Start-up Lab was established in Japan by Ari Horie in 2013 and is global in its reach and activities (womenstartuplab.com 2016).

All of the case examples of the tech initiatives and networks described above are only a fraction of the existing activities, websites, NGOs, hybrid organizations, and companies offering services, training and education especially for girls and for women. Gendering in ICT, and increasing the interest and number of girls in coding and programming have all become an interesting global wave that has the ability to transform the branch and create new innovative companies and corporations, NGOs and networks. The success models of men are desirable: "The next Steve Jobs may well be a woman. The reason I say that is for so much of the world, women are an untapped resource", (Sculley 2014). The vision announcement on the webpage of Little Miss Geek – as mentioned earlier – follows in the same spirit: "To inspire young women to become tech pioneers – we want the next Mark Zuckerberg to be female", (theempathybusiness.co.uk 2016). Several similar or resembling articulations of this message echo in the media and also on the webpages of the companies, organizations and networks presented above, emphasizing the transformative power of women, when their time comes.

4.3   GIRLS CHANGING CODES?

The narratives of the activities described above on the websites of the organizations, companies and networks follow a surprisingly similar pattern: the narrative emphasizes an exceptional individual who started a company, organization, network or collaboration project and has become aware of the gender imbalance. The corrective measures lie in the spread of knowledge and information among individuals (i.e. information for girls and young women), and adding their individual abilities to overcome the masculinity of the field, with the help of various tools. The power of education seeps through the narratives written on the webpages, and the empowering effects of the given training in coding and ICT skills are seen as a solution to the gender imbalance of the field and even more generally.

The textual strategies in the visited websites seem to follow a dual strategy in their presentations of women and men (girls vs. boys). The webpages distance women and men as agencies – or persons – from each other (following the corporate feminism idea), but at the same time, through presentations, women and men are bound together (with competences differently defined; with boys/men dominating the field, stereotypes of geeks, nerds or homosocial networks) as groups. An interesting tension arises from the celebration of the unique, individual abilities of girls/women and individual capabilities of each girl/young woman, and the simultaneous presentation of the idea that all women/girls are globally alike and in the same position vis á vis men as a group, and all find remedy through their willingness to learn coding and programming.

The strong focus on individual abilities and capabilities, and a fluid femininity with an unspecified but otherwise expressed age group preference are the common characteristics of the webpages dealing with women, technology and self-help. The Women's Start-up Lab's founder accelerator program articulates this omnipotence of women eloquently: "Don't just accelerate your start-up. Transform yourself as a founder!" (womenstartu plab.com/accel.html). Indeed, apart from one organization, TechRepublic (techrepublic.com 2016), which also delivers several other technology-related topics on its webpages, most of the webpages of the organizations, networks and companies discussed above emphasize the individuality, and individual abilities and capabilities and access to the world through programming. The issue of gender is about ability and flexibility and their new manifestations through programming and start-up activities.

The rhetoric of a justification for the need of girls in the tech world, in coding and in technology fields exemplifies the strong aim of "fixing the problem". By opening up coding through education to girls and to women, the problem of having so few women in the business will change. Some researchers discuss 'corporate feminism' (e.g. Williams 2013; Losse 2013; Williams et al. 2012) in this context. The corporate feminism is interesting because it relates to highly individualized notions of individual capabilities: for example it may be assumed that there are no boundaries for an individual, once the problem is fixed. Rather, the answer that corporate feminism offers is to "restructure the self" as the citation example above indicates. According to Losse (2013), corporate feminism does not renovate or reform the corporations or their policies or practices. Corporate feminism in general has been argued to conform to the endless requirements of work efficiency, for example (Losse 2013; Swan 2012). The volunteering in the not-for profit organizations brings in a rather similar echo of working efficiently for the purpose of others: in many of the examples taken here, the collaborative activity is based on voluntary work, such as voluntary work in teaching. The status of the volunteers in most cases remains unclear.

The textual strategies of the webpages repeat the pattern, and polish and unify it by emphasizing the unique activity that they engage in and only seldom if at all mention other networks, companies, beneficiaries or organizations who do similar kinds of work or activity. The story line with the front figure, a successful individual is most often represented as an individual who has – with the help of others – made things happen. Very seldom on the web pages are the others who have had a role in the making or birth of the network, organization or company presented as persons of character, or even mentioned. These 'other' persons who may have formally been part of the start-up phase, or own shares or have invested in the company or organization can in some cases be found mentioned by name once or twice, but they are not presented as 'characters' or persons, e.g. with pictures. In many cases a significant part of the actual work (teaching, organizing teaching and mentoring) is done in networks and by volunteers, but the volunteers, the persons doing the volunteering work, are not introduced. It remains unclear if they are ICT students who do assisting work as part of their studies (unpaid) and get study credits for the work, or whether they do it as a way to gain experience and to put it down in their CVs, or whether they do the teaching several times, or year after year, or whether it is a hobby, etc.

The presentation of the forefront figure in most cases follows the start-up companies' logic of a one-man-show and front-figure. In most of the cases briefly described above, it is indeed a question of a one-woman-show, with the initial idea coming from one woman/girl who decides to take a leap forward and start something new. The presentations follow more of a corporate or start up-pitch style, than the style of an educational institute, referring to qualifications in business and in technology, and not in education. The interest may not lie in the national educational curriculum, but in selective measures that are thought to be able to change the world of technology, in order to increase diversity in innovations.

On the webpages of the case companies and organizations these individualized femininities and the work done through the mentoring and teaching follow relatively similar patterns. It is as if they have become assembled and inscribed into gender positions and enforced through strikingly similar kinds of repetition. Men and boys are curiously absent from the web pages of the case companies and organizations: on the webpages the ICT-field in general is introduced as male-dominated and as a problem that needs to be fixed, but boys/men are not present in addressing any of those issues, nor do they bring in any corrective measures.

Through their undoubtedly influential work the companies, organizations and networks function both as accelerators and as agents transforming the ICT-field and coming up with innovations within ICT. Many initiatives mentioned here aim to solve the dilemma and help girls to develop an interest to coding. The entrepreneurial initiatives have also attracted the interest of large corporations and states. The global corporation Oracle announced in April 2016 it would invest $200 million in direct and in-kind support for computer science education in the USA. One example of the endeavours taken by Google is the creation of 13 different coding projects for girls on their website connected to the interest girls might already have in the field. This announcement was followed by the announcement by the White House of an additional $3 million for girls' education in STEM fields with the program Let the Girls Learn (Oracle 2016). Google has earlier invested $50 million to close the tech gender gap with the Made with Code initiative (Made w/Code 2016). This was preceded with the announcement that only 17% of Google's tech employees are women (Dockterman 2014). As such Google is not alone.

The systematic efforts and enduring programs to boost the education of girls in the tech fields and coding are crucial. Ways of doing this may include taking measures which are more unconventional than the standard education curriculum includes, such as arranging field trips, hackathons and boot camps. All these may bring in new audiences and offer the experiments needed for understanding more about technology fields. Undoubtedly participation in hackathons, boot camps and coding lessons help connect girls and young women with tech jobs, and lower the threshold to technology in general. It remains to be seen, however, if this means a large-scale transformation in companies and within the field with any global effects. A large-scale transformation may follow the adoption of programming into early education curriculum, in addition to college enrolment efforts, as in many Asian and European countries. In some countries, companies, with the help of NGOs, have launched programs, such as Indian Girls Code. Equally important is the shift towards more inclusive diversity in the working culture of tech companies (Baha 2016). The gender disparity in the games industry is still striking, and teaching girls how to code may get them interested in tech as a career option and is undoubtedly one key to correct the disparity.

In Europe, the European Commission has launched several programs to get more women into digital and ICT careers. The international Girls in ICT day, the Grand Coalition for Digital Jobs (part of the Digital Agenda for Europe), and the Europe Code Week, which is a grassroots initiative to bring coding and digital literacy to everybody, all aim to renew the image of the ICT sector (Ansip 2016). In Asia, several non-profit organizations have improved women's skills, access to and control over ICT through a number of pilot projects and policies (Thas et al. 2007), but lot of work is needed to push ICT and more widely STEM and science fields as integral parts of the educational curriculum for girls, and as part of poverty reduction and an equal career option for women (Unesco 2015). Schools and teachers are considered key proponents in increasing the number of girls in engineering and technology, and minority groups are argued to need more intensive courses, in order to increase the number of girls interested in technology. Activities such as after school programs or clubs may change the direction for girls (Rivas 2013). Indeed, in some Asian countries the proportion of female graduates and researchers in science, technology and innovations is near equal to the proportion of men, but the vertical segregation still remains as it does in many other parts of the world (Unesco 2015: 85).

Even as the technology sector continues to expand, women have seen declining representation in the top jobs and even in expert jobs within the industry, despite the leading forefront figures. According to 2008 statistics from the U.S. Bureau of Labor, women held only 25% of positions in computer systems design and related services. In 2015, the share of women is approximately the same: 25% of the computing workforce was women, and less than 10% were women of colour (NCWIT 2016). In Europe, the majority of jobs for ICT specialists are held by men. The proportion of women working in ICT jobs in the labour market in the EU-28 has declined, being 18% in 2014 (Eurostat 2016).

A very different take on getting gender into the gaming industry through coding is described by Mavridou and Sloan (2013) in their article about the transformative works and computer games as a participatory culture. According to Mavridou and Sloan, in 2012 and 2013, three game hacks by the game fans were released, where the gender of the main character/s had been switched, with the respect to the original work and art, and with the desire to either play as a female lead figure or allow girls to connect with the female lead (Mavridou and Sloan 2013). Sexism in the gaming culture may have diminished but there still is a need for more female characters and lead figures. A study by Williams et al. showed that 85% of playable characters in video games are male (e.g. Kasumovic and Kuznekoff 2015; Williams et al. 2009; see also Thebauld 2015; Etzkowitz 2003).

4.4　Gendering the Internet of Things

It is pouring with rain and your friends are standing at your door, ringing your doorbell, standing on the doorstep, waiting to get in. Only you are not at home, but stuck at the airport in a different city, due to a typically delayed flight. What to do and how to get your keys to them to let them in? There are no neighbours at home for your friends to pop into while they wait, and it is really getting dark, wet, and cold.

No need to worry, you just open Siri on your iPhone, and ask her to open the door, and you will immediately see a picture of your friends entering the house, relieved after getting in from the cold rain. A smart deadbolt lock on your front door and its companion app that you have installed has allowed you to unlock your door from a great distance by simply sending a request to Siri on your iPhone or iPad. The deadbolt lock also has built-in alarm technology, which senses potential door attacks, and a subtle monitoring camera at the door sends you on-line pictures or video. Your friends get in, and can start preparing a nice dinner, which will be waiting for you when you finally get home. The menu will be a wonderful surprise.

Smart home appliances are about everyday life appliances that we use all the time: phones, remote controls, smoke detectors, but also about new opportunities made possible with the help of technological innovations that connect us to many places at the same time. Smart home appliances are also examples of the interconnection of everyday objects (the Internet of things, IoT) reflecting technological innovations that can be connected instantly. Networked technological devices also gather data, and connect the data in a multitude of ways, even in uncharted ways.

The interconnection of everyday objects, interestingly, also reflects the gendered assumptions of what – as in this specific example – home is all about, and how technology and innovations can and should be used in the home, and more broadly, what IoT technological innovations are all about. Smart home appliances are either about the pleasantries of life on the one hand, or about securing, surveillance and monitoring of life on the other hand. The innovative ways the home can be used for pleasure today are in general comparable to the ways the home was used for leisure in the twentieth century. The IoT makes new innovative connections available between home and work, between sleep and health monitoring, between cooking and shopping, and overall, as some researchers note, the boundaries of the home are expanding (Strengers 2016). The new innovative solutions and the IoT require both new types of work and technical ability from consumers. This requires monitoring and surveillance.

Common to the different facets and strands of the new economy is that the technological platforms enable the scaling of new innovations and business ideas very quickly and globally. The question of how and in what ways innovations then become consumable products and services is more complex, and requires attention. Digital platforms enable "overnight" growth, that is, the rapid growth of businesses and global recognition of new innovations. This is currently visible in the gaming business and industry and related to that, in the entertainment industries that follow and capitalize on successful games. The economies of scale, which most often are about the growth of business, also lead to increasing costs when the businesses are growing. In the technology-enabled new economy, economies of scale do not materialize as fast. Also on digital platforms the development of innovations into a product can take a lot of time.

The new economy refers not only to technology-led and technology-enabled industries and businesses using digital technology as a spring board, but also to new types of growth and network related, mediating effects on businesses. In these variations, new types of innovations are spread also easily and at a faster speed and on a wider scale than earlier. These innovations are often developed incrementally but can build up speed in various ways. Let's take the example of the history of the electric washing machines discussed in Chapter 2. When electricity was integrated in the form of an electric motor to the washing machine, the shape of the washing machine changed. However, electricity did not only change the shape and form of the machine, it also changed the spatial location of the washing machine in the home, and it changed the way laundry was done both in the laundry shops and at home. The actions of the washing machine changed gradually as well: sensor-controlled technology enabled a variety of highly automated features to appear in the washing machine. In addition, sensors, thanks to their innovative technical features, differentiate different types of washing programs, the temperature of the water, drying of the laundry and even promote the self-cleaning of the machine. The machines can connect through the IoT to each other and to cloud services.

When computers become part of everyday objects, the devices and objects change. This mean also changes in the design that needs to accommodate the technology to be used in everyday life. "Smart refrigerators", also known as "Internet refrigerators" have small sensors embedded in their structures so that the device identifies the contents of the refrigerator and notifies the operator when some products need to be replaced. As research into the IoT tells us (e.g. Sundararajan 2016) a digital grocery list may appear on the display of a smart fridge. The research thus needs to ask at what price is the milk delivered to your home and whose work changes in the process? It has already been recognized that much of the non-skilled work that currently exists will change and even disappear with the digitalization of physical work and objects (e.g. McClaren and Agyeman 2015; Chase 2015; Levitt and Glick-Schiller 2008).

Researchers call the increasing connectivity and digitalization "ubiquitous computing": the design of things is no longer one-dimensional but becomes a combination of industrial design, service design and experience design. The IoT refers to the networked interconnection of everyday objects which are often equipped with ubiquitous intelligence. The phenomenon of the IoT opens up lots of possibilities for businesses and for consumers alike. Whether the IoT becomes of general use in societies alike, remains debatable, but the IoT, by integrating and interconnecting everyday objects via embedded systems, will increase the ubiquity of the Internet (Ning and Hu 2012; Xia et al. 2012). And here we will find gender re-appearing in many forms and in many ways.

The industrial Internet of Things (IIoT) refers to the ways companies capture new forms of working. But for companies and corporations it is first and foremost a tool for finding growth in new ways of organizing activities. The former 'heavy industries' such as mining, utilities, metal industries, benefit from the new organizational logistics of the IoT. The efficiency strategy promise of the IIOT, delivered by the digital services and innovations through the product mix is huge and currently taken up by the corporations both in production industries and in services such as health care. The IIoT is based on the data mining and continuous analysis of the data from the actual machines such as aircraft engines. By crafting an imagery cloud copy of the actual machine functioning through sensor data, the deviations and possibly fatal problems can be detected in time for logistically optimized reparation to take place.

What new types of connections exist between the IoT, consumers and the platform economy? It would be blatant to note that they are endless, ranging from global market places to instantaneous information flows. The platforms transform all business logic but not all businesses are exposed to platforms equally.

The augmented reality games we used earlier in this chapter as examples do not differ from the lure of other games and gaming in general. In a similar fashion as is true for many other games, such as "Angry Birds," "Pokémon Go" is free game for download. Free downloads also allow new users to adopt the game easily. The equal possibility to download the game provides seemingly democratic consumer opportunities and the free use invites more people to use it. The free game also allows a wider spread of the game, and the number of people who learn the logic of the game also become immersed more deeply into its culture. Through these types of free alluring invitations, the games undoubtedly bring in capital by tempting people to buy extras and pay for other in-app features. Games can introduce technologically mediated social networking which can constitute a significant part of the leisure activities of individuals in modern societies. This can be exemplified through the recent launch of the "Pokémon Go" game.

During the "Pokémon Go" game's first launch week in the USA 10% of Android phone users downloaded the game. Since its launch, it has become a global phenomenon in just a few weeks in 2016. The games differ but most games launched at some point gain part of their revenue from merchandizing sales and licensing deals, such as clothing, toys and sponsorship deals. The "Angry Birds" company, Rovio, for example, gained over 30% of its total revenue from merchandizing deals other than the game itself in 2011 (Bainbridge 2014; Kuittinen 2012).

Common to these games is that the original "Angry Birds" creatures were not clearly gender labelled, but since the beginning, they have added characteristics with gendered identities. This is specifically visible in the "Angry Birds" movie made. The gendered signifiers do not necessarily make the character gendered: the original "Angry Birds" game simply had different coloured birds fighting against pigs. The colours of the birds were red, yellow, blue and there was a white bird, which could lay an egg while flying, thus referring to a hen, plus a mighty eagle which had masculine characteristics in its ability to destroy things. In 2011, the "Angry Birds" company, Rovio, started to gender its game by adding the "Valentine" version of the popular game, and in the "Angry Birds" -movie some additional feminine features, such as eyelashes and hair, were added on some of the birds.

Non-gendered games are a minority though. New games keep on reinforcing the entrenched cultural notion that heroes are male by default. The asymmetry is visible also in the ways girls and women potentially project themselves onto male characters, while boys and men do not get the similar opportunities. The analysis of the gender breakdown of the annually published games shows that as long as games continue to give us significantly more stories centred on men than on women, they will continue to reinforce the idea that female experiences are secondary to male ones (Perreault et al. 2016).

Currently in the gaming industry the new economy extends beyond the game, through gaming and its rhizome networks: a way to capitalize on this is to use the momentum of global attraction, locally, as did a museum in Arkansas which found that several "PokéStops" – places where players could catch a Pokémon creature – were located across the museum's grounds. The museum innovatively photographed all the Pokémon creatures on their grounds next to various pieces of art in their collections and blogged about the appearance of the Pokémon creatures on the museum site (www.crystalbridges.org 2016), and this same 'lure' has been reported to happen around the globe, in cafés, shops and museums alike. Accordingly, it can be argued that familiar entities from our "conventional" reality are increasingly upgraded to members of IoT.

The immateriality of the monetizing process is obvious and rests on the global scaling of the original attraction. The networked and wired reality is exemplified above through games and gaming, but the networked and wired reality constitutes the whole IoT. At the same time, the games become an epitaph of modern innovations: entangled and highly connected to the current digital technologies in a multitude of ways. The price of the technology itself does not create such high barriers to innovations as earlier, e.g. at the end of the 1990s a basic chip in a hand-held GPS receiver cost over $3,000 (Masten and Plowman 2003), and now just a fraction of that price.

Investigating gender in the innovation of games development as part of the new economy brings forward a multitude of aspects and approaches from co-production and co-design through to consumerism to active agency as game developers and designers. Games and gaming communities are embedded in larger cultural contexts. These contexts repeat and redefine cultural assumptions of gender, as shown in the gender analyses of online games (e.g. Braithwaite 2014), and in the analyses of the global industry working in gaming. An analysis of gender in games and in gaming works as a prime example of how to investigate gender and gendering in the new economy. The complexity of digital networks, alignments with culture, as well as the immateriality of game consumption, and intersectionality in the virtual economy all bring forward new challenges in breaking down the gendered aspects for the further analysis.

Innovation as a specific kind of knowledge work is highly gendered. As such, knowledge work does not differ much from the gendered divisions existing in other parts of the labor markets and expert work. Not only are the innovations controlled by the heroic male images (e.g. Wynarczyk and Marlow 2010) but they also carry masculine connotations, as shown in the case examples of girls and coding earlier in the chapter. The need for achievement and independence are associated with the image of a hero also in the research literature. This extends to many of the technology-orientated activities such as coding, which are taught in schools most often in relation to mathematics, and less in relation to logic, art or languages. Thus, being good at coding is strongly related in the curriculum to mathematical skills, and to a much lesser extent to design and art, even though the visual culture is highly present in technology and gaming. One particular feature of the start-up scene, in the high-tech sector, with a strong reliance on scalable innovations, is that growth start-ups are very typically a male operation. As a counter activity to the male culture in coding, several NGOs, corporations and states have emphasized the promotion of girls into coding activities by using different instruments, such as free courses, intensive camps and summer schools, etc. Part of this activity is entwined with for-profit activities in an interesting way, as described earlier in the book. When the IoT is related to augmented reality – the question of wearable technology becomes even more complex, for example (Kao et al. 2015; Profita et al. 2013; also Du Gay et al. 1997).

4.4.1   Case: CEO of Normal

"Meet the badass founder of @normalears." was the way Rebecca Minkoff, world famous fashion designer and highly recognized designer especially in the United States introduced Nikki Kaufman, founder and CEO of the company called Normal. Normal is the New York based producer of tailor-made headphones for your phone and music player. Kaufman, the owner and CEO of the company was catalysed by her long-standing experience with ill-fitting earphones, which were particularly irritating during sports. She was also triggered by her work experience, which included 3D printing and manufacturing in a start-up company Quirky(Alleywatch 2015; Morrell 2014).

Having worked at Quirky, Kaufman was familiar with both the 3D printing and production based on 3D printing. When she got her idea, at first she started to experiment at home with prototyping. She soon found that she needed to quit her day job at Quirky, to be able to focus on the development of her key idea for earphones and 3D printing. Her business idea was an apps that allows buyers of the apps to scan their ears. The rest of the "shopping apps for manufacturing earphones" –experience is simply to choose the preferred colour for earphones, and length and colour of the cable. The scan was then sent to Normal, which converted the scan into a 3D print. The printing and production process took place in 3 hours (Alleywatch 2015; Morrell 2014).

The interconnected nature of the contemporary innovations and business start-ups is clearly visible in the case of the company Normal. Highly dependent on the mobile phone design developments, the company will rise and decline with the designs and planning of these devices. More generally, the smart adaptations of the Internet and 3D printing, such as with the company Normal described above show the growth of new adaptations in technological development. We can ask where gender is displayed in the varying earphones, but through the individual production, gender becomes adapted to the product. The similarity in outer design does not show the possible gendered differences in the product itself (Poutanen and Kovalainen 2013, 2010).

The IoT is here with the omnipotence of digital processes and technological know-how. For example, 3D printing has been used for nearly a decade. As the printers have become more accurate and cheaper, and are able to work with a broad range of materials, the printers are increasingly used in various types of production. 3D printing has made it possible to create very complex and materially demanding, built-to-order products, and increasingly so outside merely healthcare. The aim to increase the "power of the customer" with personalized products is most likely to be a success story, as people are willing to pay extra for personalized items.

Wearable technology is another adaptation of the IoT blurring the boundary between humans and technology. The topic of wearable technology is much researched and discussed within the science and technology literature, and several adaptations of wearable technology have been produced over time, starting from smart watches and exercise wristlets. Lately, MIT Media Lab and Microsoft Research have designed a metallic tattoo-type of temporary wristlet art that can control and communicate with electronic devices, such as smartphones or laptops (Kao et al. 2015). The electrodes embedded in a very thin layer of leaf gold create a conductive surface on the skin, which then, when touched can control some commands in your electronic device such as phone or iPad. It is possible to change the slides or increase the volume of the music, for example, by swiping your arm.

That the technology changes the ways we understand the companies and innovations is not new as such (e.g. Graham and Wood 2016). What makes it revolutionary in some sense is the way in which the value added production and its organization takes place. With the growth of individualized production that is not bound to a specific space or time, nor to gender in the ways it used to be, the ways innovations are gendered also change. The much lower expenses of actual production, the highly global markets and evermore stratified consumer markets show that the previous division of new businesses being small and large corporations remaining large does not necessarily hold true anymore. Platform economy, which is discussed in more details in the next Chapter, transforms several of the existing business models. The social outcomes of using these new platform models vary and they are gendered in different ways.

REFERENCES

Aarseth, E. (2004) Genre trouble: narrativism and the art of simulation. In A. Mitchell & N. Monfort (eds.) *The First Person: New Media as Story, Performance and Game*. Cambridge, MA: MIT Press.

Abbate, J. (2012) *Recoding Gender: Women's Changing Participation in Computing*. Cambridge, MA.: MIT Press.

Adams, T., & Demaiter, E. I. (2008) Skill, education and credentials in the new economy: the case of information technology workers. *Work, Employment and Society*, 22(2): 351–362.

Alleywatch (2015) Fashion Specifically for Everyone: Customizable Products. www.alleywatch.com/2015/09/fashion-specifically-for-everyone-customiz able-products/. Retrieved 14.10.2016.

Ansip, A. (2016) Working to get more European women into digital and ICT careers. https://ec.europa.eu/commission/2014-2019/ansip/blog. #Ansipblogs. Retrieved 5.8.2016.

Arvidsson, V., & Foka, A. (2015) Digital gender: perspective, phenomena, practice. *First Monday*, 20(4–6). doi:10.5210/f.

Ashcraft, C., & Blithe, S. (2010) Women in IT: The Facts. National Center for Women & Information Technology, ncwit.org.

Bacca, J., Baldiris, S., Fabregat, R., Graf, S., & Kinshuk, (2014) Augmented reality trends in education: a systematic review of research and applications. *Educational Technology & Society*, 17(4): 133–149.

Baha, T. (2016) How generation Z females could be the answer to tech's gender diversity problem. 17 Jul, 2016. Techcrunch.com. Retrieved 20 July, 2016.

Bainbridge, J. (2014) "Gotta catch 'em all!" Pokèmon, cultural practice and object networks. *IAFOR Journal of Asian Studies*, 1(1): 1–15.

Basaanjav, U. (2016) "A Girl Move": negotiating gender and technology in chess online and offline. In J. Prescott (ed.) *Handbook of Research on Race, Gender, and the Fight for Equality*. Hershey: IGI Global. 198–212.

Bogg, J., & Prescott, J. (2014) The experiences of women working in the computer games industry: an in-depth qualitative study. In J. Prescott & J. E. McGurren (eds.) *Gender Considerations and Influence in the Digital Media and Gaming Industry*. Hersey, New York: IGI Global. 92–119.

Borgonovi, F. (2016) Video gaming and gender differences in digital and printed reading performance among 15-year-old students in 26 countries. *Journal of Adolescence*, 48 April: 45–61

Braithwaite, A. (2014) 'Seriously, get out': feminists on the forums and the War (craft) on women. *New Media & Society*, 16(5): 703–718.

British Computer Society (2016) Recipients of Ada Lovelace Medal. http://academy.bcs.org/content/lovelace-medal. Retrieved 11.11.2016.

Bryce, J., Rutter, J., & Sullivan, C. (2006) Digital Games and Gender. In J. Rutter & J. Bryce (eds.) *Understanding Digital Games*. London, New York: SAGE. 148–165.

Burns, A., & Schott, G. (2004) Heavy hero or digital dummy? Multimodal player-avatar relations in FinalFantasy 7. *Visual Communication*, 3(2): 213–233.

Burrill, D. (2008) *Die Tryin'. Videogames, Masculinity, Culture*. NYC: Peter Lang.

Carr, D. (2005) Contexts, gaming pleasures and gendered preferences. *Simulation & Gaming*, 36(4): 464–482.

Chase, R. (2015) *Peers Inc: How People and Platforms are Inventing the Collaborative Economy and Reinventing Capitalism*. New York: Public Affairs.

Code (2016) Code website. http://www.codedocumentary.com/Retrieved 12.7.2016.

Consalvo, M. (2012) Confronting toxic gamer culture: a challenge for feminist game studies scholars. *Ada: A Journal of Gender, New Media, and Technology*, 1. doi: 10.7264/N33X84KH.

Decoded.com (2016) Decoded.com -website. General information. Retrieved 21.5.2016.

Delamere, F. M., & Shaw, S. M. (2008) "They see it as a guy's game": the politics of gender in digital games. *Leisure/Loisir, Special Issue: The Politics of Popular Leisure*, 32(2): 279–302.

Dockterman, E. (2014) Google invests $50 million to close the tech gender gap. *Time*, June 19, 2014.

Dovey, J., & Kennedy, H. W. (2006) *Game Cultures: Computer Games as New Media*. Maidenhead: Open University Press.

Downey, G. (2012) The here and there of a femme cave: an autoethnographic snapshot of a contextualized girl gamer space. *Cultural Studies – Critical Methodologies*, 12(3): 235–241.

Downs, E., & Smith, S. L. (2010) Keeping abreast of hypersexuality: a video game character content analysis. *Sex Roles*, 62(11): 721–733.

Egenfeldt-Nielsen, S., Heide Smith, J., & Pajares Tosca, S. (2016) *Understanding Video Games: The Essential Introduction*. New York and London: Routledge.

Eskonen, H. (2015) Kuka on kukin Linda Liukkaan maailmassa: Mozilla, Apple ja Google palaavat eläinhahmoihin. YLE: www.yle.fi. Retrieved 1.5.2016.

Etzkowitz, H. (2003) Research groups as 'quasi-firms': the invention of the entrepreneurial university. *Research Policy*, 32: 109–121.

Eurostat (2016) ICT specialists in employment. http://ec.europa.eu/eurostat/statistics-explained/index.php/ICT_specialists_in_employment. Retrieved 2.8.2016.

Evangelho, J. (2016) How 'Pokémon GO can lure more customers to your local business. Forbes, 9 July, 2016. www.forbes.com. Retrived 21.7.2016.

Fromme, J., & Unger, A. (2012) Computer games and digital game cultures: an introduction. In J. Fromme & A. Unger (eds.) *Computer Games and New Media Cultures*. Amsterdam: Springer Ltd. 1–28.

Du Gay, P., Hall, S., James, J., Mackay, H., & Negus, K. (1997) *Doing Cultural Studies: The Story of the Sony Walkman*. London, UK: Sage.

Gigaom (2014) https://gigaom.com/2014/02/13/why-facebooks-new-custom-gender-option-matters/. Retrieved 12.4.2016.

Gigaom (2015) https://gigaom.com/2015/02/26/facebook-now-has-a-fill-in-the-blank-gender-option/. Retrieved 11.4.2016.

Gilpin, L. (2014) How 'Hour of Code' sparked a movement that could teach 100 million people to code. Techrepublic.com/http://www.techrepublic.com/article/why-fixing-techs-gender-and-racial-gaps-is-more-crucial-than-ever/. Retrieved 1.7.2016.

Ginther, D. K., & Kahn, S. (2015) Comment on 'Expectations of brilliance underlie gender distributions across academic disciplines'. *Science*, 349(6246): 391–394.

Girlsintech.org (2016) http://girlsintech.org/website. Retrieved 3.5.2016.

Girlswhocode (2016) Introduction. www.girlswhocode.com. Retrieved 12.6.2016.

GraceHopper.org (2016) Introduction to Grace Hopper Organization. www.gracehopper.org. Retrieved 12.7.2016.

Graham, M., & Wood, A. (2016) Why the digital gig economy needs co-ops and unions. *openDemocracy*. Sept 15, 2016. Retrieved 14.10.2016.

Gray, K. L. (2013) Collective organizing, individual resistance, or asshole griefers? An ethnographic analysis of women of color in Xbox Live. *Ada: A Journal of Gender, New Media, and Technology*, 2: 1–21.

Gutierrez, B., Kaatz, A., Chu, S., Ramirez, D., Samson-Samuel, C., & Carnes, M. (2014) Fair play: a video game designed to address implicit race bias through active perspective taking. *Games for Health Journal*, 3(6): 371–378.

Gwee, S., Chee, Y. S., & Tan, E. M. (2013) The role of gender in mobile game-based learning. In D. Parson (ed.) *Innovations in Mobile Educational Technologies and Applications*. Hershey: IGI Global. 254–271.

Hammerman, R., & Russell, A. L. (2015) Introduction. In R. Hammerman & A. L. Russell (eds.) *Ada's Legacy: Cultures of Computing from the Victorian to the Digital Age.* New York: Morgan & Claypool. doi: 10.1145/2809523.

Harvey, A. (2015) *Gender, Age and Digital Games in the Domestic Context.* London: Routledge.

Heeter, C., Egidio, R., Mishra, P., Winn, B., & Winn, J. (2009) Alien games. Do girls prefer games designed by gGirls?. *Games and Culture*, 4(1): 74–100.

Hsu, Y.-S., Lin, Y.-H., & Yang, B. (2017) Impact of augmented reality lessons on students' STEM interest. *Research and Practice in Technology Enhanced Learning*, 12(2). doi: 10.1186/s41039-016-0039-z

Iacovides, I., Cox, A. L., McAndrew, P., Aczel, J., & Scanlon, E. (2015) Gameplay breakdowns and breakthroughs: exploring the relationship between action, understanding, and involvement. *Human–Computer Interaction*, 30(3-4): 202–231.

Ingram, M. (2014) "Average Social Gamer Is a 43-Year-Old Woman". Gigaom. Retrieved 10 June 2015.

Israel, M., Wang, S., & Marino, M. T. (2016) A multilevel analysis of diverse learners playing life science video games: Interactions between game content, learning disability status, reading proficiency, and gender. *Journal of Research in Science Teaching*, 53(2): 324–345.

Jenkins, H., & Cassell, J. (2008) From "Quake Grrls" to "Desperate Housewives": a decade of gender and gaming. In Y. B. Kafai, C. Heeter, J. Denner, & J. Y. Sun (eds.) *Beyond Barbie[R] and Mortal Kombat: New Perspectives on Gender and Gaming.* Cambridge, MA: MIT Press.

Jenson, J., & De Castell, S. (2010) Gender, simulation and gaming: research review and redirections. *Simulation & Gaming*, 41(1): 51–71.

Jenson, J., & De Castell, S. (2014) Online games, gender and feminism. In R. Mansell & P. H. Ang (eds.) *The International Encyclopedia of Digital Communication and Society.* Hoboken, NJ: John Wiley & Sons, Inc. 1–5.

Joiner, R., Iacovides, J., Owen, M., Gavin, C., Clibbery, S., Darling, J., & Drew, B. (2011) Digital games, gender and learning in engineering: do females benefit as much as males?. *Journal of Science Education and Technology*, 20(2): 178–185.

Kahn, S., & Ginther, D. K. (2015) Are recent cohorts of women with engineering bachelors less likely to stay in engineering?. *Frontiers in Psychology*, 6(1144). doi: 10.3389/fpsyg.2015.01144.

Kao, C., Dementyev, A., Paradiso, J. A., & Schmandt, C. (2015) NailO: Fingernails as an Input Surface. MIT Media Lab. Proceedings of the 33rd Annual ACM Conference on Human Factors in Computing Systems. Pages 3015–3018. April 18-23, 2015, Seoul, Republic of Korea.

Kasumovic, M. M., & Kuznekoff, J. H. (2015) Insights into sexism: male status and performance moderates female-directed hostile and amicable behaviour. *PLOS ONE*, 10(7). doi:10.1371/journal.pone.0131613.

Kidd, D., & Turner, A. J. (2016) The #GamerGate Files: misogyny in the media. In A. Novak & I. J. El-Burki Lehigh (eds.) *Defining Identity and the Changing Scope of Culture in the Digital Age*. Hershey: IGI Global. 29–37.

Kuittinen, T. (2012) Angry Birds Spreads Wings With An Ambitious Jewelry Line. www.forbes.com. Forbes, May 18, 2012. Retrieved 11.5.2016.

Levitt, P., & Glick-Schiller, N. (2008) Conceptualizing simultaneity: a transnational social field perspective on society. In S. Khagram & P. Levitt (eds.) *The Transnational Reader*. New York: Routledge. 284–294.

Lin, H. (2008) Body, space and gendered gaming experiences. In Y. B. Kafai, C. Heeter, J. Denner, & J. Y. Sun (eds.) *Beyond Barbie[R] and Mortal Combat: New Perspectives on Gender and Gaming*. Cambridge, MA: MIT Press.

Liukas, L. (2015) *Hello Ruby*. New York: Macmillan.

Liukas, L. (2016) http://lindaliukas.tumblr.com/. Retrieved 24.7.2016.

Losse, K. (2013) Feminism's Tipping Point: Who Wins from Leaning in? Dissent, March 26, 2013, online magazine (www.dissentmagazine.org).

Lowood, H. (2006) Game studies now, history of science then. *Games and Culture*, 1(1): 78–82.

Lutz, H., Herrera Vivar, M. T., & Supik, L. (2011) Framing intersectionality: an introduction. In H. Lutz, M. T. Herrera Vivar, & L. Supik (eds.) *Framing Intersectionality. Debates on a Multi-Faceted Concept in Gender Studies*. London, New York: Routledge. 1–20.

Lykke, N. (2011) Intersectional analysis. Black box or useful critical feminist thinking technology?. In H. Lutz, M. T. Herrera Vivar, & L. Supik (eds.) *Framing Intersectionality. Debates on a Multi-Faceted Concept in Gender Studies*. London, New York: Routledge. 207–220.

Made w/Code (2016) https://www.madewithcode.com/. Retrieved 13.10.2016.

Masten, D., & Plowman, T. (2003) Digital ethnography: the next wave in understanding the consumer experience. *Design Management Journal*, 14(2): 75–81.

Mavridou, O., & Sloan, R. J. S. (2013) Playing outside of the box: transformative works and computer games as participatory culture. *Participations. Journal of Audience and Reception Studies*, 10(2): 246–259.

McCall, L. (2005) The complexity of intersectionality. *Signs: Journal of Women and Culture and Society*, 30: 1771–1800.

McClaren, D., & Agyeman, J. (2015) *Sharing Cities*. Cambridge, Massachusetts, London, England: MIT Press.

Morrell, K. (2014) Nikki Kaufman of Normal: Custom Earphones Without the Custom Price Tag. American Express Open Forum. https://www.american express.com/us/small-business/openforum/articles/nikki-kaufman-normal-custom-ear-puds-without-custom-pricetag/. Retrieved 23 July, 2016

NCWIT (2016) Statistics on IT workforce. National Center for Women & Information Technology. https://www.ncwit.org/ncwit-fact-sheet. Retrieved 15.7.2016.

Ning, H., & Hu, S. (2012) Technology classification, industry, and education for Future Internet of Things. *International Journal of Communication Systems*, 25(9): 1230–1241.

Ochsner, A., Ramirez, D., & Steinkuehler, C. (2015) Eductional Games and Outcomes. In R. Mansell & P. H. Ang (eds.) *The International Encyclopedia of Digital Communication and Society*. Hoboken, NJ: John Wiley & Sons, Inc. 1–8.

OECD (2014) *PISA2012 Results in Focus. What 15-year-olds know and what they can do and what they know*. Paris: OECD.

Olsson, T., Lagerstam, E., Kärkkäinen, T., & Vainio-Mattila-Väänänen, K. (2013) Expected user experience of mobile augmented reality services: a user study in the context of shopping centres. *Personal and Ubiquitous Computing*, 17(2): 287–304. doi: 10.1007/s00779-011-0494-x.

Oracle (2016) 'let the girls learn'-programme. http://www.oracle.com/. Retrieved 15.5.2016.

Oxford English Dictionary (2016). Oxford: Oxford University Press. Public.oed. com. Retrieved 2.8.2016.

Parmar, B. (2012) *Little Miss Geek: Bridging the Gap Between Girls and Technology*. London: Lady Geek Ltd.

Parmar, B. (2015) The Most (and Least) Empathetic Companies. Harward Business Review, Nov. 27, 2015. hrb.org/2015/11/2015-empathy-index. Retrieved 24.4.2016.

Perreault, M. F., Perreault, G. P., Jenkins, J., & Morrison, A. (2016) Depictions of Female Protagonists in Digital Games: A Narrative Analysis of 2013 DICE Award-Winning Digital Games. *Games and Culture*, 11: 1–18. doi: 10.1177/1555412016679584.

Pierce, C. (2015) Foreword. In A. Harvey (eds.) 2015 *Gender, Age and Digital Games in the Domestic Context*. London: Routledge. 1–3.

Poutanen, S., & Kovalainen, A. (2010) Critical theory. In A. Mills, G. Durepos, & E. Wiebe (eds.) *Encyclopedia of Case Study Research*. Thousand Oaks, London, New Delhi: Sage Publications. 261–265.

Poutanen, S., & Kovalainen, A. (2013) Gendering innovation process in an industrial plant – revisiting tokenism, gender and innovation. *International Journal of Gender and Entrepreneurship*, 5(3): 257–274.

Prescott, J., & McGurren, J. E. (2014) Introduction. In J. Prescott & J. E. McGurren (eds.) *Gender Considerations and Influence in the Digital Media and Gaming Industry*. Hersey, New York: IGI Global.

Profita, H., Clawson, J., Gilliland, S., Zeagler, C., Starner, T., Budd, J., & Yi-Luenen, D. E. (2013) *Don't mind me touching my wrist: a case study of interacting with on-body technology in public*. Proceedings of ISWC'13, Sept. 9-12, 2013. Zurich, Switzerland.

Rivas, L. (2013) How the DIY girls afterschool program offers a path to technology careers for underserved girls. InformIT. www.informit.com/articles/arti cle.aspx?p=2145967. Retrieved 23.4.2016.

Royse, P., Lee, J., Undrahbyan, J., & Consalvo, M. (2007) Women and games: technologies of the gendered self. *New Media & Society*, 9(4): 555–576.

Sculley, J. (2014) Conference talk, Web Summit Conference 2014.Techworld. com, http://www.techworld.com/news/startups/ex-apple-ceo-john-sculley-next-steve-jobs-may-well-be-woman-3584213/Retrieved 20.4.2015.

Searle, K. A., & Kafai, Y. B. (2012) Beyond freedom of movement boys play in a tween virtual world. *Games and Culture*, 7(4): 281–304.

Shaw, A. (2010) What is video game culture? Cultural studies and game studies. *Games and Culture*, 6(5): 403–424.

Sigurdadottir, H., Skevik, T. O., Ekker, K., & Godejord, B. J. (2015) *Gender Differences in Perceiving Digital Game-Based Learning: Back to Square one?* In European conference on games based learning. Reading: Academic Conferences International Ltd. 489–496.

Strengers, Y. (2016) Envisioning the smart home: reimagining a smart energy future. In S. Pink, E. Ardèvol, & D. Lanzeni (eds.) *Digital Materialities: Design and Anthropology*. London: Bloomsbury. 61–76.

Sundararajan, A. (2016) *The Sharing Economy: The End of Employment and the Rise of Crowd-Based Capitalism*. Cambridge, Massachusetts, London, England: The MIT Press.

Svenigsson, M. (2012) 'Pity There's So Few Girls!' attitudes to female participation in a Swedish gaming context. In J. Fromme & A. Unger (eds.) *Computer Games and New Media Cultures. A Handbook of Digital Games Studies*. New York, London: Springer. 425–442.

Swan, E. (2012) Cleaning up? Transnational corporate femininity and dirty work in magazine culture. In R. Simpson, N. Slutskaya, P. Lewis, & H. Höpfl (eds.) *Dirty Work: Concepts and Identities*. Houndsmills, Basingstoke: Palgrave Macmillan. 182–202.

Techrepublic.com (2016). Homepage. Retrieved 3.6.2016.

Thas, A. M. K., Ramilo, C. G., & Cinco, C. (2007) *Gender and ICT. United National Development Programme. Asia-Pacific Development Information Programme*. New Delhi: Elsevier.

Thebauld, S. (2015) Business as plan B: institutional foundations of gender inequality in entrepreneurship across 24 industrialized countries. *Administrative Science Quarterly*, 260(4): 671–711.

Thierer, A. (2014) *Permissionless Innovation: The Continuing Case for Comprehensive Technological Freedom*. Mercatus Center: George Mason University.

Turkle, S. (1995) *Life on the Screen. Identity in the Age of the Internet*. New York: Simon & Schuster

Turkle, S. (2011) *Alone together. Why we expect more from technology and less from each other.* New York: Basic Books.

Unesco (2015) *A Complex Formula: Girls and Women in Science, Technology, Engineering and Mathematics in Asia.* Paris: UNESCO. Seoul: KWDI.

Voorhees, G. (2014) Neoliberal masculinity: the government of play and masculinity in e-sports. In R. A. Brookey & T. P. Oates (eds.) *Playing to Win: Sports, Video Games and the Culture of Play.* Bloomington, IN: Indiana University Press. 63–91.

Voorhees, G. (2016) Daddy issues: constructions of fatherhood in the last of us and bioShock infinite. *Ada: A Journal of Gender, New Media and Technology,* doi: 10.7264/N3Z60MBN.

Williams, C. (2013) The glass escalator, revisited: gender inequality in neoliberal times. *Gender & Society,* 27(5): 609–629.

Williams, D., Martins, N., Consalvo, M., & Ivory, J. D. (2009) The virtual census: representations of gender, race and age in video games. *New Media & Society,* 11(5): 815–834

Williams, C. L., Muller, C., & Kilanski, K. (2012) Gendered organizations in the new economy. *Gender & Society,* 26(4): 549–573.

Winn, J., & Heeter, C. (2009) Gaming, gender, and time: who makes time to play?. *Sex Roles,* 61(1): 1–13.

Women's coding collective (2016) Introduction. http://thewc.co. Retrieved 23.6.2016.

Womenstartuplab.com (2016) Women's Start-up Lab, About us. http://www. womenstartuplab.com. Retrieved 5.3.2016.

Wright, C. (2016) A Brief History of Mobile Games: In the beginning, there was Snake. www.pocketgamer.biz. March 14th, 2016. Retrieved 10.12.2016.

Wynarczyk, P., & Marlow, S. (2010) Introduction. P. Wynarczyk & S. Marlow (eds.) *Innovating Women: Contributions to Technological Advancement.* Bingley, UK: Emerald.

Xia, F., Yang, L. T., Wang, L., & Vinel, A. (2012) Internet of things. *International Journal of Communication Systems,* 25(9): 1101–1102. doi: 10.1002/dac.2417.

www.crystalbridges.org (2016). Retrieved 4.8.2016.

www.ifgirlsrantheworld.com (2016). Retrieved 17.5.2016.

www.itu.int (2017) Retrieved 5.4.2017.

www.gameslearningsociety.org (2016). Retrieved 15.6.2016.

www.theempathybusiness.co.uk (2016) Retrieved 6.6.2016.

www.us.macmillan.com (2016). Retrieved 1.6.2016.

CHAPTER 5

Creative Work and Gender

5.1   SPANNING THE BOUNDARIES OF CREATIVE WORK

The creative economy is often used as an umbrella concept. It usually includes the contributions of those who are in creative occupations outside of the creative industries too, as well as those who are employed in the creative industries i.e. in artistic and cultural fields and jobs. The idea of care seldom fits into the thinking and discussions about the creative work. But with the new economy also care is changing. In fact, care is one critical nexus through which we can problematize the new economy and its construction.

The contemporary changes in the work and organization of work towards the project based, flexible and short-term workplaces put more emphasis and responsibility to individuals for finding creative solutions in getting a job and keeping in, but also creativity in crafting the job in ways it responds to own skills and capabilities. Creativity at work thus becomes understood in several new ways, when the entrepreneurial risk becomes part and parcel of the job.

One of the researchers of the new cultural social class, Richard Florida, has noted that, contrary to the commonly held beliefs, creative work does not only belong to and originate from the start-up companies, research laboratories or artists' studios. It can be located in factories and factory workers' varying solutions in their work tasks. Florida's studies of highly performing factories in the USA of the 1980s and 1990s served as a springboard for his more general theory of the creative class (Florida 2012). For Florida, the ways the factory workers came up with basic improvements in productivity and performance represented creativity at work, and intellectual capital. There are many studies of creativity and the role of knowledge in the factory, which extend the idea of creativity to the tasks and jobs at hand, from the actual occupational groups (see e.g. Oakley 2006, 2011; Gordon 2000; Zuboff 1989; Kenney and Florida 1990). Mostly, the understanding of the individuals' worth and the view into the craftsmanship kind of knowledge is at the basis of creativity at work (Sennett 2008). With this type of definition, creative work widens beyond the so-called creative industries and creative economy as such, into continuing involvement and the desire to do 'good work', as Sennett argues.

Several authors have recently argued that media and creative industries are at the forefront of economic and technological changes and developments globally (e.g. Neff 2016; Oakley 2011; Ashcraft and Blithe 2010). This may be partly due to the fact that the global economy is becoming increasingly "communicative", that is, the economic value of services, products and corporations is found in communication, in the ability to mediate the symbolic values to consumers, and in the visual and deeply cultural aspects of the corporation represented through communication (see Poutanen et al. 2016). Indeed, as the economic value is increasingly immaterial and mediated through visual and cultural aspects, the creativity of all work that provides the economic value becomes understood as being more expansive than ordinary artistic work. On the other hand, the gaming industry could claim the same outcome: artistic work is truly an integral part of the game design and gaming industry. The gaming industries and related sectors lean on artistic work of coders and their interpretative capabilities. The artistic work of coders thus reflects and mediates for its part the culture. The web pioneers, designers, web maintainers, bloggers and game designers are all doing this type of new creative work. These jobs and tasks in traditional sense were not even existing prior the digitalization and the Internet.

But creativity at work is much more than a creative job title, and, more importantly, creativity is not determined and restricted by job descriptions. We may well ask how creative work works, and what do people do when they do creative work. The answers most probably vary highly. Indeed, this is the case if creative work is understood to take place in the highly skilled and precise work of surgeons and welders as well as in the highly flexible work of coders, some examples to mention. Florida notes that roughly one third of the current workforce in the developed world can justifiably be classified as members of the so-called creative class (Florida 2002a). Statistics classify most often several industries that can collectively be called the creative industries. The definition of creative industries is not consistent between countries and has also changed nationally, as in the UK and in Australia (e.g. Higgs et al. 2007a, 2007b; Pagan et al. 2008). The occupation classifications included within the creative industries thus vary as well. Indeed, the creative industries have been more changeable over the last decade than many fields of high research intensity, such as biotechnology.

The idea of creative class bundles the creative individuals and ties them together through creative industries and creative economy, which for its part gives rise to the importance of place. Place has indeed importance and increasingly so, as some cities fail to grow ultimately due to lack of tolerance and openness, and regional economic development is based on layers of social and economic fabric, of creativity and abundance of possibilities for creative identity validation (see also Chapter 3). In the contemporary world, the creative class does not form a unified group or 'class' in a traditional sense. The digital technologies have enabled the creation of communities and networks of interest, irrespective of whether there is any common denominator for those involved. There is a need to look at the relationship of gender and creative work in relation to wider canvass.

The measurement of how important the creative industries and sectors are in the new economy is indeed an enigma: how to dissect the creative element from other work tasks and activities, especially when work is intangible and abstract. The ways gender is entangled with creativity and creative jobs adds to the complexity of the question. If we take 'the easy way out' and look at the statistics, the creative jobs in statistics are considered to include only those working in the creative industries themselves, and who may either be in creative occupations or in other roles at the creative industries, e.g. in finance (Department for Culture 2016). Thus, creative manager Alice – introduced later in this Chapter – would count as a person who works in creative industry also according to statistics, as she works in advertising. Even if Alice is not herself in a creative occupation and does not do creative work because she works with budgets and funding, she is classified in statistics as a creative industry worker.

It is estimated that the number of people who work in creative industries will greatly increase in the coming years, and the key workforce will consist of highly educated and specialized persons. In the UK, for example, the number of people working in the creative sector in the economy outnumbers the skilled workforce with a STEM educational background. The creative industries' workers and STEM-educated workers hold one thing in common. This is that the skilled work tasks and complex jobs of both creative workers and STEM workers tend to be less susceptible to automation (Sleeman 2016). Research work is often considered creative. Problem solving in science is in fact used as one measure of scientific creativity (Stumpf 1995; Ochse 1990).

Many social scientists point out that the complexity underlying creativity at work often blurs the idea of how and in what ways the job fulfils the idea of creativity. The most common definition for what is a creative job starts paradoxically by excluding what creative jobs are not: when the degree of control over the work tasks is very low and the intrinsic rewards are non-existent, the job is not considered creative (e.g. Kalleberg 2011). Intrinsic rewards reflect the ways individuals find that their skills, capabilities and knowledge are in use. Very often, the ability to set the pace of one's work is related to the intrinsic value of work. Some researchers relate the ability to use one's skills in the job to the quality aspect. Several associate creative jobs with high-paying, professional and knowledge intensive fields that require specialized education (e.g. Florida 2002b) and individuals with sufficient cultural and social capital, and work experience (McGivern et al. 2015; Tweedie 2013).

The attempts to measure the value added of the cultural industries and occupations, work and jobs included various classifications have been developed, particularly for the use of the policy makers within governments (e.g. Higgs et al. 2007a). The creative industry groupings and segments derived from statistics of occupations and industries cover often essential parts, but not all of the creative economy, as classifications do not capture the quick changes and developments taking place with mobile platforms and platform economies, and the ways individuals adapt to these forms of work (e.g. McMullin and Dryburgh 2011; McDowell 2008a; Hassan and Purser 2007; Hochschild 1983).

Neff et al. (2005) studied the ways in which the entrepreneurial labour becomes understood as creative work in the new economy through the cultural quality of 'cool', autonomy and also foreshortened careers. The media and entertainment industry – as part of the creative economy – is one of the fastest-growing fields in the creative industries. Also the fashion industry, although much smaller than the new media industry, has always thrived in vibrant regions and urban centres, where both symbolic capital (Lash 2002), 'cool' and attractive jobs (Gill 2002) and creative class (Florida 2002a) reside and create positive incentives. According to the results of Neff et al., the status within new media industry is not conferred by job title or pay, as the best jobs in the new economy, especially in new media, are both creative and entrepreneurial (Neff et al. 2005). The differences between creative, entrepreneurial work and corporate, non-entrepreneurial work are often associated with risks attached to the work, and with privileges, especially in creative new-media jobs. The risks concerning creative work relate to short-termism and qualifications that may not be transferable elsewhere.

The new attitudes towards job security are present in studies that deal with the contemporary workforce in creative industries (e.g. Brinkley 2016). A growing feature of the new economy in general is the growth of the 'free professions', which is visible in the growing number of self-employed and own-account workers. For example, in the UK 2% of the total workforce reports themselves as freelancers, and the increase in the number of freelancers resonates with the overall rise in self-employment (Brinkley 2016). According to the studies, for many the question of free-lancing is more about how to get paid and earn a living than how to work as a lifestyle choice (e.g. Vallas 2011).

All the examples of growth in the self-employment in creative jobs indicate how the nature of work has changed (Kovalainen 1995). The question of freelancing in creative industries widens to all industries. The study on freelancing in the European Union (EU) focused on own account self-employed workers and shows that across EU the share of 'IPros' – independent professionals – rose from c. 3% in 2004 to 4.1% in 2013. The US labour markets have functioned rather differently. The comparable element between countries is missing, because the definition of freelancers is different in USA in comparison to UK and EU. However, 2016 estimates by Katz and Krueger suggest that there has been a remarkable increase in contingent and alternative work between 2005 and 2015 (Katz and Krueger 2016; Kalleberg 2011). According to these estimates, the net increase in employment in the USA over the past decade has been solely in alternative forms of employment (mainly temporary labour and contract workers). This concerns especially migrant workers (Levitt and Jaworsky 2007).

Moving back to creative industries, a new attitude that is visible in several studies among the creative industries' workers is reflected also in Neff's study on venture labour (Neff 2016). As a concept venture labour means the way in which people act like entrepreneurs and carry some of the risks of the company as if they own the shares or have the ownership. This kind of labour embodies the contemporary time where new media workers' entrepreneurial behaviour reflects cultural shifts in contemporary workplaces. It also reflects the new attitudes towards job security. Employees are increasingly asked to throw themselves into work and use their social and intellectual capital, empathy and passion to keep up the corporation spirit. The insecurity of job turns into an individualized portfolio thinking. Gill's study of the freelance new media workers in six European countries involved work as digital animation, web design and broadcasting and digital arts and design. The study showed that while the work was considered as 'cool', it seldom offered permanency. In addition, the gendered positions and inequality in terms of salary differences between women and men, and career advancement were perhaps even stronger than in the 'traditional' media (Gill 2002).

The ways the creative work is recognized and defined, has much to do with artistic, commercial or cultural values of societies, but perhaps even more with temporal factors. Creativity at work was differently understood in the 1970s and 1980s than in 2010 and 2017. It can be noted that cultural industries focus on the commercialization of expressive values of music, television, radio, publishing, computer games and film, these being the key dimensions. The precariousness of commercialization also makes the jobs vulnerable but there are also other trends that shape the jobs in creative sectors. The digital platforms discussed earlier in this book also shape the micro-businesses and self-employment in the digital and creative sectors (Brinkley 2016; Hathaway 2015; Hathaway and Muro 2016).

The new working model of short-termism and temporary work has become one of the signals of the dynamism and innovations in the economy. The digitally enabled gig economy is not only about people working in short-termism, freelance or gig work but also about nesting new economic activities and small-scale entrepreneurship globally through Etsy, Airbnb, Uber and Taskrabbit. The question of creativity is not the same for all those platforms: for Etsy, the classical entrepreneurial small-scale activity and community building is highly present in its platform business model. For Uber, the business model is very different from Etsy, as the Uber driver cannot set the price for the service but is free to do own activities in-between the drives. The spectrum of digital labour and business platforms is wide, and other types of digital platforms further widen the scope of the platform economies.

For creative industries, platforms enable wider reach for audiences and customers and lucrative markets, as showed earlier. Gig economy, for its part, allows short-term contracting, innovative ensembles and non-employer firms to flourish, but not without costs. The production of attractive and informative platforms requires new types of professional skills and capabilities. For example, vlogging through YouTube can at best be highly profitable, and vlogs are definitively part of the creative work created and re-created through digital platforms. The question remains, what types of qualifications do these gig economy works contain, and do these jobs contain any larger employee-driven innovations (e.g. Hoyrup 2012).

The direct financial value of the creative industries can be measured in terms of its economic contribution. Globally, the gross value added of the creative industries has grown and the parts of creative arts such as music, performing and visual arts have grown in importance. The largest constituent parts of the creative industries, such as IT, software, and computer services, advertising and marketing have become both value adding and employing industries, and have grown quicker in comparison to other industrial fields, in the UK as well as in the USA. The share of the creative industries of the UK total gross value added is over 5%, and the creative economy as a whole accounts for over 8% of the UK economy (Department for Culture, Media and Sport 2016; also Bakhshi et al. 2013).

Today, globally the cultural and creative industries are major drivers of the modern economies of developed countries. However, in developing countries also the role and emphasis of creative industries has grown exponentially. Creative sector influences income generation, job creation and export earnings of nations. Indeed, it is estimated that worldwide, the revenues of the cultural and creative industries exceed those of telecom services. It is estimated that the top three employers globally, both as art forms and as businesses, are the visual arts, books, and music (EY 2015). Currently, creative industries develop across borders, cross fertilizing and reinvigorating the cultural fabric of societies. The importance of intangible sector is undisputed. Still, little is known of the creative work as such. It has indeed been argued that this is because the economic weight of cultural and creative industries in mature and emerging economies is only partially described, often misunderstood, and generally undervalued (Lampel and Germain 2016; EY 2015). One aspect is that creative industries necessarily function as other industries and business sectors in their managerial and organizational practices and structures. Research has shown that these also vary among and within the creative industries (Foster et al. 2011).

It is difficult to give a simple and straightforward definition of what creative work entails, and who are the people working in creative jobs, and the exact share of the volume of creative jobs, industries, and the economic sector. One reason for the vagueness in exact numbers is due to different ways of understanding what and how creativity is part and parcel of many jobs, occupations and business logics. One way of understanding what creative work entails is to look at the whole creative economy and its importance in the whole economy. The creative economy covers all the creative industries and services, well beyond the arts and cultural goods and services. Currently research and development are often included in the creative economy as well. Some researchers discuss not creative work but creative occupations, while other researchers emphasize the innovative side of creative work that is not bound to occupations classified as creative.

In most countries, ministries and state departments have become conscious of the need to define and think about creativity, both in education and in the ways it determines and defines certain occupations and jobs. This has led to estimate the number of people who work in creative occupations. The estimates are often either based on occupations which are directly connected to creative industry, or individuals who are self-employed, or affiliated with the creative end results. In the UK this estimate is 6% of total employment that is around 2 million people. This figure, based on a definition by the official UK statistics, includes advertising and marketing, architecture, crafts, museums, music, publishing, design in relation to products, graphics and fashion, film, TV, etc.

The size of the creative industry is directly connected to the size of the consumer markets. Even if USA has been dominating the global markets, diversification has already taken place, paradoxically due to global Internet. For example, the Asia-Pacific region is home to 47% of the global online population, and Asia-Pacific countries are among the most connected in the world. All these create possibilities and opportunities for creative work and creative industries. Indeed, the Asia-Pacific region leads the global gaming industry with 47.5% of the global market, and contributes in a major way (82% in 2014) to the growth of the global games market, benefiting from the rise of online gaming (EY 2015). It has been argued that technological changes set the boundaries between the "elements of the world" (e.g. Hesmondhalgh 2013: 223), such as public and private, care and capital accumulation. Given the wide array of creative work and jobs in contemporary societies the ways creativity relates to innovations in jobs and work is highly complex, ambivalent and even contested. How the creative work manifests itself in technological platforms, how gender becomes displayed and how broad systemic changes become actualized are discussed below.

5.1.1 Case Study: Everyday Technology and the New Economy in Alice's Life

Alice works as one of the senior advertising/creative managers in a multinational IT corporation. Her work is high-pressure and performance oriented management work where strong analytical and communication skills are required. When asked what type of work she does, Alice talks more about budgets than ideas. She does not directly describe her own work in advertising as creative work. She thinks that her work experience in high tech and training in management, advertising and accounting has led her to a position where she mainly deals with creativity, but she increasingly feels that her own training and occupation is not inherently creative. The pressures of performance exist increasingly in her daily work. However, Alice can be regarded as a member of the creative class in society, as defined by Florida (2002a, 2002b, 2014). In her work, Alice's ability to influence others is in high demand, and that creates stability amid creative chaos. She feels however that she is at times more an accounting director than a creative director. She enjoys the atmosphere of her unit, and thinks of it as being very creative.

Alice prefers to walk to her office instead of using the tube. This means one hour longer travelling to work daily, and also means that she leaves her home for the office often at 6 am, but she enjoys her daily walking exercise. She no longer needs to carry papers back and forth, only her mobile devices. Versatile technology solutions, such as the cloud platform her company uses, makes her work social, and she feels connected to her team and corporation and to her key clients. Cloud service her company uses enables her work to be mobile, flexible, distant – and always on.

On a personal level, Alice also wears other tech devices than her mobile, which most of the day is in her hand or close-by. Irrespective of her location, Alice checks her fitness wristband regularly during the working day, and sometimes discreetly also in meetings, if she is not in charge of running the meeting. The wristband Alice wears measures her heartbeat and sleep rhythm, counts the steps she takes, the number of stairs she climbs, and the calories she burns during the day and night. Alice has noticed that fitness wristbands are increasingly used by young men, but less commonly by women in her office. Nevertheless, she is determined to keep it on in order to keep track and remain fit. Being fit, energetic, youthful and not fat in her industry is a non-spoken must, as the whole industry and especially her own work environment is highly competitive. She has already decided to purchase a newer version of the wristband, with more functions.

The functions Alice follows are part of her daily routines. Her activity rates, heartbeat, quality of sleep and exercise progress tell her not only about her fitness but also about the need to focus and to exercise. Her exercising is indeed sometimes irregular and delayed due to intensive and long working days, and also to the fact that she visits and takes care of her elderly mother during most weekends. Luckily, the wristband alarms her when the exercises get too irregular. The device synchronizes the data with her other fitness devices, such as the scale which has a wireless Internet connection. The scale's mobile app in Alice's phone helps her to track the changes in weight and in body mass index. The synchronization of all data allows Alice to follow up and analyse the progress and changes with her laptop. The GPS of the device, as well as the GPS in her phone, help her to map new, possibly interesting routes for her walks and runs. Overall, the technology and its adaptations help Alice with her personal work-life balance and with almost everything else, but not with the most wanted thing, namely to have more hours in the day.

When we revisit the snapshot of Alice's life and reflect on the work Alice does, we can ask whether she is part of this creative economy or not. Alice, who works in advertising section of an IT corporation, thinks of herself as a creative manager with work tasks closer to management and financing than creation of new products and services. Simultaneously she notes that she works with innovation people and sees them as an integral part of her own work. Predicting that they have their future in work is partly what Alice does. We can justifiably ask whether her work is creative work. Alice hesitates herself to say this, but after some thought she answers straightaway "yes", despite the managerial side of her work. Alice sees that the products, their design and the overall image is group work and she is part of that group who, in the process, develops the end product further. However, in statistics the managerial work Alice does is not classified as part of the creative economy. Even if Alice works in close collaboration with the product and service design and advertising, she is not directly creating products or services. Alice's job is to strategically oversee, supervise, and in general pull the strings when needed and make sure resources are in line with the tasks at hand. Without her and her job, product design unit might suffer when financial cuts and rearrangements are introduced. Alice's work is closely aligned with the work of the designers and creative work in general. Making a clear distinction and separation between the enabling input of Alice's work and creative input of designers would not be possible. The work Alice does is one example of the difficulties in drawing the boundaries for creative/non-creative work in innovative industries such as the IT world.

Our example, and the scale and scope of two critical and often overlapping parts – creative companies and creative occupations – give an understanding of how wide-ranging work in the creative economy is, and how closely it relates to the overall economy. There is a general need for more data and stronger indicators on the role of culture for the development of societies and economies, and for the development of innovations.

The creative enterprises and industries include also non-profit cultural organizations and commercial businesses that produce and distribute products in which the creative content defines their market position and audiences. In addition, individuals who are self-employed are often part of the creative economy. In the statistics, however, only those directly involved in creative activity are included. Classifying those employees who are responsible for routine and non-creative functions and jobs for every successful creative enterprise is more complex. Alice, our advertising manager is typically in a job where her belonging to creative class can be questioned. Defining the borders permanently is difficult.

A creative enterprise is often defined as a company for which the primary or major value of its products or services is rooted in its emotional and aesthetic appeal to the customer or the markets. A creative occupation is often defined as a job in either a creative industry or non-creative industry, and in which the work itself is inherently creative or artistic. The creative economy encompasses both of these groups, which also overlap, as the case below shows.

5.1.2 Case Study: Gigle – Creative Art Production Platform

Gigle Inc. was established by Inkeri Borgman and Janne Matilainen in 2015. The company platform and the forthcoming app by Gigle searches and delivers the best artists and performers for any events and festivities. The origins of the start-up were in the paternity leave of Mr. Matilainen, who, prior they had their twins, worked as game designer in a game company in Finland. Having developed successful games for the company over several years, he was ready to try something new. Ms. Borgman has both education and work experience in culture production. As Mr. Matilainen was on paternity leave the couple started discussing and developing their joint business idea: bringing culture to work places, weddings, birthday parties and other festivities. Their ambitious aim is to launch a platform that offers services not only in Finland but elsewhere. They wish to become a global player and platform in the culture production. Gigle mediates cultural activities and brings together the culture experts, artists and those who wish to use those services, consume culture and arrange cultural events or other events that would use artistic production as part of the event. Gigle is in the launch phase, and as with many start-ups, the finances and contents are in constant flux. For the couple, Gigle as a platform is an intermediary organization but also a market where shared interests meet. The business model and the ways to handle the complexity of the cultural platform are currently under construction.

Creative work today gets even more entangled and increasingly complex to define, as our examples show. Work most often includes elements that are not as such 'creative' but which are necessary for the creative outcome of the work process, such as budgeting, financing and project planning. More so than earlier, creativity and innovation is also group and teamwork building on previous innovations, achievements and experiences.

It is often quite straightforwardly assumed that many of the changes in the society or in the economy can be traced back to globalization and technological development, and thus, these two – globalization and technological development – become containers for all kinds of changes. Analogously, the new economy has come to epitomize myriad things, any new technology, ICT, software work and immateriality of economy, all leading to similarities in labour, work and personhood relations, as shown in both the case descriptions above.

5.2 CARE AND TECHNOLOGICAL INNOVATIONS

The ways in which the new economy and technological platforms function in care work and in healthcare more generally are numerous and the whole issue is a multi-layered phenomenon with complex involved history and expanding future. Partly due to that, our focus in this chapter is on questions of whether we can distinguish the new economy within the field of care, what does it possibly entail, how technology changes care work and whether the distinction is relevant in the field of care and well-ness, as technologies change care and care work in multitudes of ways.

The intersections between new types of transnational work, migrating workers, gender and new types of global dependencies and interdepen-dencies have been discussed widely in relation to changing care work and care chains (e.g. Huang et al. 2012; Yeates 2008, 2012; Dahl et al. 2011; Isaksen 2011; McDowell 2008b). Care as work is shaped by the social and political institutions and contexts, especially when we talk about global care chains that have become reality in the global economy (Wallerstein 2014; Tronto 2011; Walby 2009).

Care has always been part of human activity: formal care as profession and informal care as part of family or kin relations, as act of love and as obligation as well. Care and caregiving are gendered as practices and gender is very much also about caring. Providing care is often considered as an activity that somehow requires feminine qualities, and femininity is indeed often considered to have a caring nature. As formal and paid work, care was feminized due to its close relations to traditional idea of women's role in caring and biological nurturing relations. Even when care activities are paid, the work remains undervalued: care occupations have tradition-ally been not only women's work, but have also involved lower wages than non–care occupations. Thus, one of the key elements in care is in how many ways the gender is attached to care and how the possible effects of the technology influence on care and on gender. In the following, we will look at the general effects of technologies into the gendered care as work, in relation to the new economy.

Common aspect to all care in new economy as well as in the old is the labour intensity of the work, and the timely aspect of it: care cannot be postponed or transferred but it is situated and time-bound person-centred work. Technical solutions and innovations are helpers in solving the care puzzle, but do not take away the core idea of the care: humanity. The labour intensity of care effects to the ways care is organized. One solution to the increasing cost of care is technology. The technology has become part of care and care work in many ways. The presence of technology ranges from devices and diagnostics of healthcare, drug development and medical treatments to the ways care is understood both highly personal care. In personal care work, the effects of technology are often mediated. The technological aspects in healthcare, in all forms of care and in wellness are strongly gendered by nature: irrespective of the form or type of care – be it then healthcare, social care or actual work and jobs in these fields – gender has a role to play in each of these aspects.

Technology is crucial for care, and technology helps not only those who are cared for but also carers. Technology is great enabler, as State of Caring UK 2011 report showed: an overwhelming majority (72%) of the carers who used technology noted that it gave them greater peace of mind, irrespective of whether the question was of technology assisted medication, health monitoring at home, or safety (www.carersuk.org 2016).

At the macro and population level, of all the socio-economic factors that influence health and wellbeing, gender is particularly significant. While women at the population level have in general lower mortality rates than men in Western countries (e.g. Annandale 2014), women also experience greater morbidity at the population level, and they are in general over-represented in healthcare statistics as care recipients (Dahl et al. 2011). Gender in general has a significant effect on health differences and disparities in care. These differences are not, however, simply differences between men and women. Gender-related factors also lead to significant divisions within each gender, thus reflecting various other societal factors producing and reproducing inequalities. Gender analysis is crucial to distinguish between biological causes and social explanations for the health differentials between men and women, and to understand that these gaps are outcomes of the unequal social relations between men and women, and not merely due to consequences of biology.

Gender analysis in healthcare and in medicine often provides information on the deviations in diagnosis, treatment and prevention. There is a well-documented history of male bias in medical diagnostics, such as cardiovascular disease and use of aspirin, done with over 22,000 male physicians only, and multiple risk analysis on coronary heart disease, also done on a male sample only (Rosser 1994; Schiebinger 2000). Even if the diagnostics in medicine are based on gender sensitive analyses and take women into account in the medical testing and diagnostics, there are still several blind spots. One of the key points of departure is how new gender sensitive research questions are introduced, and how much time is given to revamp the methods and trends in the natural sciences. Development of gender sensitive and gender specific indicators for illnesses is indeed part of the gender analysis (e.g. Lin and L'Orange 2012; Ettorre and Kingdon 2012).

At the same time, gender-specific knowledge and the gender bias currently embedded in medical research has become more visible, even if the phenomenon itself is well recognized as an 'old' problem. Questions such as women's lack of control over their bodies and inequalities in health status between men and women are among the major concerns that require research and political intervention and change. These aspects are important, and are also related to gender in care in general.

The ways in which the new economy and technological platforms function in care work and in healthcare more generally are numerous, and the whole issue is a multi-layered phenomenon with a complex history and an expanding future. Partly due to that, our focus in this chapter is on questions of whether we can distinguish the new economy within the field of care, what it might entail, how technology changes care work, and whether the distinction is relevant in the field of care and wellness as technologies change care and care work in many ways.

Why is care as gendered work interesting in relation to the new economy and technological developments and innovations? The care work requires the embodied presence of the carer and the person who is cared for. Care as work and as labour carries in itself power as well as dependency and vulnerability. The new developments in care work, often discussed as manifestations of the new economy, such as short-termism of work contracts, fragmented work practices and transnational gendered care may all for their part put forward their strains to care as labour and as work. Care has already become global labour through the global care chains, being partly market-based and commoditized and partly fragmented work through contracts, but these facts are not enough to change the nature of the required embodied presence in the care work and the embodied location of the care. Care work still takes place in personal contact between carer and cared-for, but this embodied relationship carries in itself, apart from the labour intensity, also larger set of social relations within which the caring is done, perceived and assessed by others.

Healthcare services are rapidly growing and needed globally, especially in the global North, much due growing elderly population. Globalization processes overall, and more specifically the worldwide demographics bias in the global North all set new demands for healthcare and for social care. In addition, the higher standard of living has fuelled the number of innovations in healthcare technologies and also for its part increased the global migration of professionals in the healthcare sector. Contemporary aspect of the global trends in the new health economy is the global migration of care workers (e.g. Esplen 2009; Kingma 2006; Khadria 2007). The health professional mobility and migration is one aspect of this phenomenon. The health professional migration also reflects the deep socio-economic and political divisions that are often discussed as global North-South–divide. This unequal division of possibilities, despite the high level of education, has led to the migration, the patterns of which for healthcare professionals go along the North-South–divide. While affluent households may have the option of paying for care, in poorer households intense care needs are often met only by women's informal work input. These divisions and income differences reign throughout the globe.

Healthcare services are globally also becoming a highly productive sector of the world economy, despite the labour intensity of the sector. Globalization processes and the worldwide increase in demand for personal care and healthcare have together not only fuelled the spread of technologies globally, but also started a new, different global trend. The global care chains mean the migration of expert healthcare labour force, resulting also the global migration of healthcare workers beyond the professional groups such as doctors and nurses. Thus, not only medical doctors and nurses but also child-minders, elderly care workers and cleaners have become globally mobile workforce. According to OECD statistics, the migration of healthcare professionals has dramatically increased in scale mostly due to the liberalization of markets and changes in population dynamics over the past two decades (OECD 2007; Eckenwiler 2009). Active recruitment in many countries has additionally increased the number of those migrating: for example, international nurse recruitment and migration based on systematic recruitments have been increasing in the last decade. Recent trends show an increase in the migration of nurses from developing countries to developed countries, resulting in a worldwide shortage of nurses (Levitt and Jaworsky 2007; Dywili et al. 2013).

The migration patterns are difficult to picture accurately but various studies indicate a pattern that is characterized by global migration from global South, that is, from low- and middle-income countries to global North, high-income countries in North America and Western Europe.

The OECD study showed that nearly all European OECD countries have increasingly relied on recruiting health workers from abroad to fill their shortages already ten years ago (OECD 2007; Favell 2007). There are currently several factors contributing to and causing the global migration of care work and care workers, apart from the factors above. Research has shown that better remuneration for work abroad, professional advancement prospects and better career opportunities, a safer working environment and a better quality of life are the main reasons for migration (Dywili et al. 2013; Isaksen 2011). Additionally, one such factor is the ways the governments structure their policies around the questions of care. Countries with strong welfare state support and provisions for care have less demand for private care than in countries with weak welfare state structure. However, the strong welfare state can invite migration of the skilled care workers due to other reasons, such as relatively high salaries and insufficient number of care work professionals available at the national labour markets.

It has been estimated that the global migration of healthcare work professionals will continue and increase exponentially with the new technology helping to alleviate language and work culture problems. For example, in the USA, the number of overseas-educated medical doctors has grown in seven years from 2002 to 2009 by 70%, and in most OECD countries, the share of foreign-trained medical doctors has been increasing in recent years as well (OECD Observer 2010). The mobility figures tell also about the excellent human capital through training and mobility upwards in societal ranks across the borders.

Several European countries, for example UK and Spain, have established bilateral agreements to recruit health workers. These shifts of labour can have both positive and negative implications, depending on the country's role in the migratory scheme. Indeed, even in strong welfare states where the support for the dependent and their caregivers is professionalized, or formalized, there is more demand for migrant workers than where the schemes for care are more informal (e.g. Cuban 2013; Dahl et al. 2011).

In order to say something about the phenomenon, it is essential to know who those healthcare professionals are who migrate and where do they migrate to. According to recent large-scale global study 57% of the respondents migrated to a country where the same language is spoken. Of them, 33% migrated to neighbouring countries and 21% migrated to former colonizing countries, even if the language was different to mother tongue. Women medical doctors and nurses, and male and female nurses migrated to neighbouring countries, nurses and older and highly educated healthcare workers to former colonizing countries and highly educated health workers and medical doctors to countries that have a language match (De Vries et al. 2016). It seems that the high education and qualifications do not stop the migration, but on the contrary, they give good possibilities for migration, as the education and the language skills are assets at the global care markets.

Professional healthcare mobility and migration are highly gendered, to the extent that gender has in care work and professions given rise to the global care chains. The term global care chain was first introduced by Hochschild (2001) to remark how migrant care workers replace and fill in the care deficits borne out of two main reasons: the increasing number of women at the paid labour markets and the inabilities of the states to respond to this with public measures. Care work is not piece-meal assembly line factory work, but highly interpersonal, temporally bound and intimate work with sick, frail or fragile people. Thus becoming accommodated and acclimatized to the receiving society or locality seems to be one key element of being able to do the demanding care work.

Parallel to the ideas of work in the new economy, the questions of transnationalism and globalization relate to the ways work is presently understood and organized globally. The question of what is counted as promotable new economy and what counts as old economy, and how transnationalism relates to these two forms of thinking the gendered work, becomes even more pertinent, given the fact, that for example, it is estimated that over 50% of the global workforce can be defined as working in the informal sector (Williams 2011). While we can claim that highly interdependent national economies are increasingly becoming a single market economy, at least in Europe, for highly skilled and highly educated professionals, other types of gender-differentiated and local labour markets in care are also emerging (Williams and Round 2008; Williams et al. 2012). These types of work are often assumed to be located among the less skilled or poorly skilled, among women and even in the shadow economy (Cuban 2013). Much of this work is informal care work that exists in all societies. Thus, care work, transnationalism and globalization are not reducible to one issue only, but relate to several things such as locality and dependencies of different types (extended familial ties, relatives). The ways in which technology is related to and entangled with care today, is thus also layered and not reducible to one meaning only. Technologies may change the practices but not the orientations to work. Indeed, in empirical interview studies, care professionals themselves do not consider their work ethics or practices as dependent on the contractual nature of the work, nor on the type of work contract they have (Kovalainen and Österberg-Högstedt 2013).

To critique the rigid division into 'old' and 'new' economies, the relationship between old and new needs to be more carefully carved out and contextualized. As referred earlier, the 'new' most often refers to the digital and software sectors and parts of the economy, in contrast to manufacturing, piece work and labour intensity. The inherent assumption is that all work that is governed by and done within these sectors is part of that new, or the old, economy. But, and our thesis is, nothing in society is untouched by any other; the old and new economies blend into each other. The Filipino or African woman who has migrated to the USA to work as a 'shadow mother' (Macdonald 2011) or household cleaner /gardener uses technologies provided by the new economy when sending money through the SWIFT system from the States back home to a local bank or to family through mobile banking system. The old economy jobs and work and the contents – gardening, cleaning, caring or cooking for children and spending time with them – have not changed, but they have submerged in themselves the devices and services provided by the new economy.

Through the care work, the idea of the work and its radical change in the new economy can indeed be problematized. Care in the new economy does not entail qualitatively different elements than the work done in the old economy, as shown in the analysis by Poutanen and Kovalainen (2014). Yet, the change in the institutional contexts shapes also the forms and ways of working in the care. Their analysis shows how the care constantly requires presence and bodily work, and how the technologies of care and power are in differing ways inherently present in those work activities. It can be argued that the gendered care work resists and challenges the often made strong division and classification into the old and new economies (Poutanen and Kovalainen 2014).

Interestingly to the new economy thesis, some large corporations have adopted the 'old' informality as part of their inherent corporate strategy (Williams 2011: 3). As an example of this, salaries even in many global corporations in many countries can still be paid in two lump sums, declared pay and undeclared 'envelope wages' (Williams 2011). This example of wages paid in two formats – formal and informal, not hidden – shows distinctively not only how the old economy inherently exists within the new economy, but perhaps more importantly, how the domestication processes of globalization spread and function. In this specific case of wages paid in two formats, it also shows how culture is engraving the contours of the new economy (Sweet and Meiksins 2012) that take different shapes in different contexts.

The arguments of the new economy and gender often state that the dynamics of the new economy have provided gendered opportunities and constraints, for women as well as for men (e.g. McDowell 2008a, also McDowell 2008b). It is true that gender relates to and is constituted by and within the new economy in so many ways it becomes difficult to disentangle the various threads and discussions, so no general overview is given here either. Understood most broadly from a feminist point of view, the new economy as part of globalization has facilitated and enabled new spaces, institutions and shared rhetoric where universality, through globalization, in terms of human rights has become one justificatory principle (e.g. Walby 2002, 2009). The major advances in the feminist theorization of globalization – covering some aspects of the new economy as well – have been mainly twofold: to systematize the gender effects of globalization (Tronto 2011; Naples 2008), and to intersect globalization theories and gender theories with contextualized interrelationships (Bair 2010; Yeates 2012).

One example of the complex contextualized interrelationships is shown by Isaksen's analysis of migrant care workers in Italy (Isaksen 2011). The tradition in Sicily, Italy has been that care workers were recruited from neighbouring villages away from the direct local community. Today according to Isaksen, these 'neighbourhood wives' are replaced with 'neighbouring country' wives. Contemporary patterns of informal care mean today the crossing of national borders instead of villages and payment related to different foreign currencies. In Central Europe, the traditional 'neighbourhood wives' taking care of the inferior and intimate care work came from lower social classes (Isaksen 2011). Migrant care workers, 'new neighbourhood wives' have in most countries become partially integrated into the local economy and work force, partially living far from their homeland country. Global markets thus bring in a new care economy in terms of migrant care workers (OECD Observer 2010), and with that, an institutionalization of temporality and distance.

5.2.1   Case Study: Lisa, Alice's Mother

Alice's mother Lisa suffers from several illnesses that restrict her ability to be fully independent and manage without some home help and some assisted care. She lives in her own house where she has been living most of her adult life since her marriage. Her house is impractical, as it is large with several stairs and so she needs help to live and to maintain the house. Currently she is attended by a home care helper two to three times a week. The home care and care professionals from a care company help Lisa with cleaning and maintenance of the household, grocery shopping and other everyday activities once or twice a week. That is the amount of help Alice can afford to purchase to her mother.

Even if Lisa still manages perfectly well at home, the family discussion about Lisa moving to an assisted apartment is ongoing. So far technology helps her to maintain her independent living, and she prefers that. Lisa also continuously wears a wristband, but her wristband is a very different type and for very different purposes than her daughter's. Lisa's wristband monitors her movements and alarms if no movement takes place. Alice has had a security camera, door alarm and a surveillance system installed at Lisa's house. The security system is not necessarily needed as the neighbourhood is safe, but it gives Lisa a feeling of safety.

Alice can no longer think of her life – or her mother's life – without the technological devices and technological help such as the tracking and monitoring systems, which help her in many ways. The tracking devices help her to keep in contact with her mother from a distance, but also help her to focus on her own health and wellbeing. Similarly, Alice cannot think of her life – or her mother's life – without some sort of wearable technology such as the wristband. The technology helps her to cope with her own everyday life. Technology does not, however, provide company to her mother in the evenings. But Skype and Facetime calls help to stay connected. Overall, the new technology helps Alice to organize her mother's everyday life and wellbeing. The care company contract concerning the services Alice and Lisa purchase for Lisa's welfare also states the increasing use of technology in services such as in the monitoring of Lisa's wellbeing and needs. With the use of technology, the care company is able to reduce the number of individual visits it makes.

Lisa has several types of devices at her home that support her independent living at home, despite her chronic rheumatism. The devices range from remote alarm system to smartphone and far-field voice control. As

Lisa spends a lot of time at home, she has learned to enjoy the company of online radio, for example, and she often uses the Internet also through the voice control devices, either in the living room or on her cell phone. The voice control device answers her questions of the phone numbers, weather around the globe, and it even reads her audiobooks and the news. With the device, she gets the weather reports and opening hours of the local shops, etc. In fact, there is little the voice controlled far-field system does not do, but they do make mistakes – or Lisa does when asking one thing and getting an answer to an entirely different matter. Hence, it is about learning to specify questions and being simple. The devices seem to develop and the current remote voice control Lisa has can even control the lights of the house and the thermostats, but those qualities Lisa has not tried yet. For Lisa, it is not an option that her phone or voice controlled device would keep her the only company, though and so far, the devices have not yet replaced a human contact for her.

The snapshots above of Alice's and Lisa's everyday lives narrate a more general picture and rather usual mix of informal and formal care arrangements in a many Western countries: adult children keep contact and even do the distant-care for their elderly parents. Even if Alice is not considered directly as family caregiver, as her mother, Lisa, manages mostly on her own, she sometimes feels stressed due to her mother's care and support requirements. Several research reports evidence how the adult children and family caregivers cite higher levels of perceived stress, and most often the caregivers are women. Also, the difficulty to find time to care for oneself, the feeling of social isolation and the lack of work-life balance, result in a negative impact to their overall emotional well-being (e.g. Tao and McRoy 2015). The research carried out by Carers UK showed that majority (72%) of the family members who use technology in care mentioned that it gave them 'a greater peace of mind'. However, technology is not available for everybody in caring work in similar manner: 2011 Survey by State of Caring showed 46% of carers in the UK did not know how to access supportive technological solutions, even if these solutions would have been financially or otherwise available (www.carersuk.org 2016). The short glimpses to Alice's and Lisa's lives exemplify also how closely technology in general, the various uses of technology, its innovative adaptations and the huge progress in the ways the technology is available in special, is entangled with wellbeing and also with modern care in Western societies. The new economy includes several aspects that disrupt, change and rearrange the previous arrangements and established understandings in those economic fields where gender is most involved, especially in care work and in creative work. But does technology wipe away the gendered nature of care work, and to what extent technology can change the gendered patterns of informal care? This chapter has discussed the new economy and the transformation of the gendered care in general. And as the snapshot of Alice's and Lisa's lives showed, these are not so far from each other.

5.3   Hybridization of Care Work

In everyday life the differentiation in the gendering of the technology is highly visible in the wearable gadgets and technical devices attached to the clothing. The fitness tracking which started with devices that originally measured the steps taken, has since developed into measuring the intensity of activities and numerous bodily functions. The bursts of exercises, for example, are much more important for the sports activities than just measurement of length of the exercise. Rather soon in the technology development, gender differentiation was taken into account, for example by embedding the tracking devices into clothes such as sports bra. The devices – no matter where they are in your sports clothing, usually measure the heart rate and breathing rhythm and prompt you accordingly. The algorithm in the wearable technology lets you know when you are training in your best zone by measuring your exercise based on the cadence, steps, pace, stride length and calories burned, for example.

Would you like to check from your phone how you feel? Do you wish to have an algorithm to know when your next period starts, or do you possibly have the symptoms of PMS coming instead of ordinary headache? The mood changes in relation to menstruation are no longer viewed as a natural period of hormonal imbalance but they have become medicalized as a bunch of symptoms of premenstrual tension (PMT) and thus treatable and traceable with the help of apps. These examples are first and foremost about gathering data of oneself with help variety of sensors, apps and other technology. But will you really learn to know yourself through a 'datafied and quantified self'? (e.g. Neff and Nafus 2016; Nafus and Sherman 2014; Profita et al. 2013; Topol 2012).

Health monitoring and self-monitoring, with the help of wearable technology, and the mapping of the individual's bodily functions with the data gathered with wearable technology are two recent examples of the contemporary, data-driven trends in the healthcare and wellness self-monitoring. The global markets for health self-monitoring technologies have grown exponentially, to the extent that both the technological devices and the apps are seen the as the biggest shake-up in the personal healthcare, health monitoring and prevention of several illnesses as the market opportunities for new health self-monitoring products and technologies are globally growing. Indeed, the digital health has emerged as one key dimension of contemporary healthcare policy and delivery in many countries.

While the telemedicine has generally become a standard part of the healthcare policy, breaking the national boundaries in new ways, the new applications of digital services in tele-medicine, have spread exponentially. Telemedicine in general involves the use of digital and other technologies to encourage patients to self-monitor their medical conditions at home (e.g. Oudshoorn 2012; Barta and Neff 2015). Another part of the tele-medicine is the expert interpretation done of medical cases through the Internet (remote experts interpreting x-rays or other medical images). As one example of the new trends in human-machine interfaces and sensory modalities, the biofeedback trend has grown exponentially (Swan 2012).

The 'connected care' integrates all parts of the healthcare system, from patients and carers to doctors and hospitals, and as it has been claimed, even to insurers and government with real-time networks. A study on the spread of telemedicine estimated that 61% of US healthcare organizations have adopted telemedicine solutions such as remote monitoring and diagnosis consultations (HIMSS, 2016). The three biggest perceived barriers to connected technology adoption in healthcare according to professionals are the cost of devices, the privacy concerns (data security), and health system bureaucracy (FHI 2016).

Biofeedback uses the information from technological devices and/or sensors that receive and monitor information (feedback) of the body (bio). The growth of biofeedback use has during the last few years grown into new intensity with several new wearable technologies and apps that enable the monitoring. Several applications have been around quite a long time to measure health, fitness and other types of data, ranging from blood pressure and calories intake to blood glucose and alcohol content. All of these are advertised as new ways to learn about yourself through the data you collect (Ramirez 2013). By measuring the heart rate, breathing levels and, for example, the indicators of stress, biofeedback technology has developed psychologically conscious video games with stress-control techniques that help to fight tension. It is another matter if video games are the best way to fight the tension, but increasingly self-help materials in video and in YouTube are used for alleviating stress.

For many, such as for Alice, the activity trackers and with them the monitored data have become one key dimension in maintaining health and wellbeing and it also includes fitness data, exercise data, food entry, sleep data etc. Advocates of digital health and wellness technologies argue that individuals as consumers are an integral part of the digitalization that is about to occur or has already occurred in healthcare, and that they should play an active role in 'digitizing' their bodies and its functions (e.g. Topol 2012; Swan 2012). These devices also represent the Internet of things: monitoring devices are most often synced wirelessly with a computer or smartphone for long-term data tracking. The downsides of the activity trackers are several and equally, several of them are about complex ownership of the data, permission to gather digital data from apps, and more generally, the privacy matters. Issues such as apps for some activity trackers not only transmit personal data, but also private address lists to servers on the Internet without notifying or asking the user's permission.

The answers to questions such the one addressing how and by whom the vast amounts of data produced in the health and care technology is used, include ideas and business start-up possibilities. Most apps programs ask – when logging in and selecting any data type for review using program access – whether the company can read, maintain and storage your data. After granting permission, a visual display of the data you have gathered yourself is available to yourself and to any other you permit access to your data. This data can be highly personal, or it may also refer to population averages, in order to locate and allow for comparisons to take place.

Another dimension of the technology becoming part of the everyday life, health and social care is the wearable technology. Wearable technology means technology that is integrated to the clothes and clothing materials. Intelligent clothing is clothes and clothing materials where additional tracking and data mediating, gathering and creating materials have become part of the clothing and clothes (e.g. Malmivaara 2009). The boost to do research on wearable technology was supported by DARPA (Defense Advance Research Projects Agency) and research started with highly technical emphasis, with development of wearable computers, minimization, virtual reality and augmented reality (McCann and Bryson 2009). It was only after the beginning of 2000s when design of clothing became part of the wearable technology developments, with many different types of experiments following. Today the digital tailoring and conductive fibres have expanded the wearable tech into diversification and individualization which both emphasize the potentiality of technology. Technology can take many forms and be part of the everyday life in many ways, indeed, wearable technology strands range from protection (patient care, prevention care for elderly, and for high-impact sports) to prohibition of injuries, support, active self-monitoring and health maintenance and lifestyle support (active population).

In recent years, the increasing social impact and interconnection of care work with care markets, carers (actors), and institutions have been all understood under the broad concept of social and healthcare and its transformation or different forms. Moving from globalization to the new economy, the connection is close. The new economy and its workings presuppose globalization, and globalization for its part feeds into the new economy, which seems to be inherently assembled as part of the globalizing processes. So the assemblages of the new economy/old economy, gender, care work and globalization become interesting to explore further, as we have done above albeit briefly. We have argued, through the intersection of gender, care work and economy that the clear-cut divisions such as the new economy/old economy do not work in complex analysis.

More importantly, care work as embodied labour challenges the basis of the claimed foundation of the new economy vis-à-vis the old economy. The assumed broken relationship between the personhood and labour does not hold true as principled difference between the two, but accommodates both in the embodied presence of care work/labour. In this chapter, we have shown through examples that the rigid division between the old economy and the new economy is challenged and rejected by the gendered care work and labour. By focusing on the question of personhood and the new economy, we also wish to evoke the more general question of whether we can make such as clear and analytical distinction between the 'old' and the 'new' economies, as often assumed? We have argued for a different solution, through the powerful example of care work.

The question we have interrogated relates more widely to methodological questions and challenges. The growing interrelatedness of nation states is visible in the shifts that have taken place through the last decade from isolated containers to supranational actors (e.g. the idea of the European Community) and from fortresses to horizontal interdependencies and new power constellations (e.g. the recent global financial crisis and the Eurozone crisis within Europe's euro-using countries, such as Greece and Spain) that call for specificities and defined social contexts.

These specificities and defined social contexts should not assume any given causality between global changes and locally bound activities as such. This discourse of changing institutions is in alignment with the growing critique on the nation-state container theory of society, which seeks to move beyond the view of the states as container units of analysis. The analysis of the processes of the adoption of common policies and 'travelling policies' show us different kinds of domestication processes taking place. Can we therefore talk about the same, global or European policy, for example (Morel et al. 2012)? The simultaneity of engagement in activities across and beyond nation states, for example, as in migration, sets additional methodological and theoretical challenges for research on globalization, transnationalism and the new economy.

Several authors have argued that specific entrepreneurial behaviour of workers relies on 'venture labour', not on actual entrepreneurial activities within organization (e.g. Neff 2016). This behaviour reflects a broader transformation in society in which economic risk shifts away from collective responsibility and toward individual responsibility. The platforms do not employ, nor do they own the means for production. They become assembled as initiators and igniters for individual economic activities. The paradox in these activities is that they are formatted activities, following the script of contract with no entrepreneurial freedom existing within a script. For that reason, entrepreneurial and innovation related vocabulary is not suitable in describing the work in the labour platforms.

In the new economy, risk and reward takes the place of job loyalty throughout the labour markets (Sennett 1998). Some researchers label the management of the risks of contemporary work as 'venture labor' (Neff 2016). What exactly is venture labour? Neff defines it as the investment of time, energy, human capital and other personal resources that employees make in the companies where they work (Neff 2016). The ways the workers adapt to new technologies and mold the work, the emerging posts and positions in the new economy into their own is highly contingent, and adaptable, resembling the entrepreneurial values of non-entrepreneurs. The possible risk that related to the short-termism and fixed positions was framed in a study by Neff – not in terms of financial gains from working in the start-ups but in terms of desire to work in creative environment. The ways people take up the idea of ownership – without having any formal ownership – of their workplaces can range from off-hours commitment, as in many workplaces, to social networks promoting the companies on free time.

The concept of venture labour leaves out the downside of the venture capital, which is, losing the capital. As for venture funding, there is a possibility to insure the capital against particular risks and losses, and for gains the margins are usually calculable. But for labour as 'capital' the venturing does not function in a similar manner as for other types of capital: the deep commitment to the corporation does not necessarily bring in better possibilities for staying within the corporation in the economic turmoil or recession (Sennett 2006). The transferable skills and capabilities can offer better chances in the future labour markets.

One of the central concepts for creative work is a deep commitment to the actual work. Indeed, work identity is used as a concept to refer to a work-based self-concept, constituted and constructed in the making and defined in the literature in multiple ways (e.g. Martin and Wajcman 2004). As a construct, work identity overlaps with other constructs used in the studies of creative work, such as occupational identity so that they are often used interchangeably (e.g. Levine 2010). How to get a grasp of work identity? As an elusive concept, work identity is difficult to measure. The articulations of work identity take place in the processes of meaning-making within organizations, when individuals contextualize and situate their occupational skills relative to the skills and capabilities of their co-workers and the organization and its needs. The questions of identity and identity construction connect to work through several layers and activities. For many, job and career building are intertwined with creative work and more so, self-worth (e.g. Beech 2008). Having to re-manufacture your worth every day at work is exhausting and hollowing out experience in the venture labour times.

The idea of identity work builds on Goffman's (1967) concept of social encounters. Social encounters allow people to enact who they are. A wider question then is how people's identities become meaningful to themselves and others. Questions about willing hybrids' professional identities lead them to challenge and disrupt institutionalized professionalism, and use and integrate professionalism and managerialism, creating more legitimate hybrid professionalism in their managerial context.

References

Annandale, E. (2014) *The Sociology of Health and Medicine: A Critical Introduction*. Cambridge: Polity Press.

Ashcraft, C., & Blithe, S. (2010) Women in IT: The Facts. National Center for Women & Information Technology, ncwit.org.

Bair, J. (2010) On difference and capital: gender and the globalization of production. *Signs*, 36(1): 203–226.

Bakhshi, H., Freeman, A., & Higgs, P. (2013) *A Dynamic Mapping of the UK's Creative Industries*. London: Nesta.

Barta, K., & Neff, G. (2015) Technologies for sharing: lessons from quantified self about the political economy of platforms. *Information, Communication & Society*, doi: 10.1080/1369118X.2015.1118520.

Beech, N. (2008) On the nature of dialogic identity work. *Organization*, 15(1): 51–74.

Brinkley, I. (2016) In search of the Gig Economy. The Work Foundation, August 2016. www.workfoundation.com. Retrieved 28.9.2016.

Cuban, S. (2013) *Deskilling Migrant Women in the Global Care Industry*. Basingstoke: Palgrave.

Dahl, H. M., Keränen, M., & Kovalainen, A. (2011) Introduction. In H. M. Dahl, M. Keränen, & A. Kovalainen (eds.) *Europeanization, Care and Gender: Global Complexities*: 1–19.Basingstoke: Palgrave.

De Vries, D. H., Steinmetz, S., & Tijdens, K. G. (2016) Does migration 'pay off' for foreign-born migrant health workers? An exploratory analysis using the global WageIndicator dataset. Human Resources for Health, 14(1): 40–51.

Department for Culture, Media and Sport (2016) *Creative industries economic estimates January 2016 – key findings*. www.gov.uk/government/publications/creative-industries-economic-estimates-january-2016. Retrieved 12.8.2016.

Dywili, L., Bonner, A., & O'Brien, L. (2013) Why do nurses migrate? – A review of recent literature. *Journal of Nursing Management*, 21(3): 511–520.

Eckenwiler, L. A. (2009) Care worker migration and transnational justice. *Public Health Ethics*, 2: 171–183.

Esplen, E. (2009) *Gender and Care. Overview Report*. Bridge, IDS. http://www.bridge.ids.ac.uk/. Retrieved 15.6.2016.

Ettorre, E., & Kingdon, C. (2012) Reproductive regimes: governing gendered bodies. In E. Kuhlmann & E. Annandale (eds.) *The Palgrave Handbook of Gender and Healthcare*. Houndmills, Basingstoke and London: Palgrave Macmillan. 162–177.

EY (2015) *Cultural Times. The First Global Map of Cultural and Creative Industries*. Brussels: EYGM Ltd.

Favell, A. (2007) Rebooting migration theory. Interdisciplinarity, globality and postdisciplinarity in migration studies. In C. Brettell & J. Hollifield (eds.) *Migration Theory: Talking Across Disciplines*. 2nd ed. London: Routledge. 259–278.

FHI (2016) Future Health Index 2016. The capacity to care. Measuring perceptions of accessibility and integration of healthcare systems, and adoption of connected healthcare. Retrieved 17.9.2016 from: https://www.futurehealthindex.com/report/2016/.

Florida, R. (2002a) *The Rise of the Creative Class*. New York: Basic Books.

Florida, R. (2002b) Bohemia and economic geography. *Journal of Economic Geography*, 2: 55–71.

Florida, R. (2012) *The Rise of the Creative Class, Revisited*. New York: Basic Books.

Florida, R. (2014) Europe in the creative age, revisited. *Demos Quarterly*, 1, 2014–2014. Winter. http://quarterly.demos.co.uk/article/issue-1/europe-in-the-creative-age-revisited-7/. Retrieved 11.12.2016.

Foster, P., Borgatti, S. P., & Jones, C. (2011) Gatekeeper search and selection strategies: relational and network governance in a cultural market. *Poetics*, 39 (4): 247–265.

Gill, R. (2002) Cool, creative and egalitarian? Exploring gender in project-based new media work in Europe. *Information, Communication & Society*, 5(1): 70–89.

Goffman, E. (1967) *Interaction Ritual. Essays on Face-to-Face Behavior*. New York: Ancor Books.

Gordon, J. (2000) The Hands-on, Logged-on Worker. *Forbes*, 30 October, 2000. Pp. 136–142.

Hassan, R., & Purser, R. E. (2007) Introduction. In R. Hassan & R. E. Purser (eds.) *24/7. Time and Temporality in the Network Society*. Stanford: Stanford University Press. 1–23.

Hathaway, I. (2015) The Gig Economy is real if you know where to look. *Harward Business Review*. Aug. 15, 2015.

Hathaway, I., & Muro, M. (2016) Tracking the gig economy: new numbers. Research report. October 13, 2016. Brookings Institute.

Hesmondhalgh, D. (2013) *The Cultural Industries*. London, UK, Thousand Oaks, CA: SAGE.

HIMSS Analytics (2016) *2016 Telemedicine Study*. April 2016. Retrieved from: http://www.himssanalytics.org/research/hmss-analytics-essentials-brief-2016-telemedicine-study. 12.10.2016.

Higgs, P., Cunningham, S., & Pagan, J., (2007a) *Australia's creative economy: Basic evidence on size, growth, income and employment*. Technical Report. Brisbane: Faculty Research Office, CCI. http://eprints.qut.edu.au/archive/00008241/.

Higgs, P., Cunningham, S., & Pagan, J., (2007b) *Australia's creative economy: Definitions of the segments and sectors*. Technical Report. Brisbane: Faculty Research Office, CCI. http://eprints.qut.edu.au/archive/00008242/.

Hochschild, A. R. (1983) *The Managed Heart: Commercialization of Human Feeling*. Berkeley: University of California Press.

Hochschild, A. R. (2001) The nanny chain. *The American Prospect*, 11(4): 32–36.

Hoyrup, S. (2012) Employee-driven Innovation: A new phenomenon, concept and mode of innovation. In S. Hoyrup, K. Möller, M. Bonnafous-Boucher, C. Hasse, & M. Lotz (eds.) *Employee-Driven Innovation. A New Approach*. London: Palgrave Macmillan UK.

Huang, S., Thang, -L.-L., & Toyota, M. (2012) Transnational mobilities for care: rethinking the dynamics of care in Asia. *Global Networks*, 12(2): 129–134.

Isaksen, L. W. (2011) Gendering the stranger: nomadic care workers in Europe. A Polish – Italian example. In H. M. Dahl, M. Keränen, & A. Kovalainen (eds.) *Europeanization, Care and Gender: Global complexities.* Basingstoke: Palgrave.

Kalleberg, A. (2011) *Good Jobs, Bad Jobs: The Rise of Polarized and Precarious Employment Systems in the United States, 1970s–2000s.* New York: Russell Sage Foundation.

Katz, L. F., & Krueger, A. B. (2016) The Rise and Nature of Alternative Work Arrangements in the United States, 1995-2015. Working paper. Princeton University.

Kenney, M., & Florida, R. (1990) *Beyond Mass Production: The Japanese System and Its Transfer to the United States.* New York: Oxford University Press.

Khadria, B. (2007) International nurse recruitment in India. *Health Services Research*, 42(3): 1429–1436.

Kingma, M. (2006) *Nurses on the Move: Migration and the Global Health Care Economy.* Ithaca: Cornell University Press.

Kovalainen, A. (1995) *At the Margins of the Economy: Women's Self-Employment in Finland, 1960–1990.* Ashgate: Avebury.

Kovalainen, A., & Österberg-Högstedt, J. (2013) Entrepreneurship within social and health care: a question of identity, gender and professionalism. *International Journal of Gender and Entrepreneurship*, 5(1): 17–35.

Lampel, J., & Germain, O. (2016) Creative industries as hubs of new organizational and business practices. *Journal of Business Research*, 69(2016): 2327–2333.

Lash, S. (2002) *Critique of Information.* London, Thousand Oaks, CA, New Delhi: Sage Publications.

Levine, D. P. (2010) *Object Relations, Work, and the Self.* London, New York: Routledge.

Levitt, P., & Jaworsky, B. N. (2007) Transnational migration studies: past developments and future trends. *American Review of Sociology*, 33: 129–156.

Lin, V., & L'Orange, H. (2012) Gender sensitive indicators for healthcare. In E. Kuhlmann & E. Annandale (eds.) *Palgrave Handbook of Gender and Healthcare.* London: Palgrave Macmillan. 92–110.

Malmivaara, M. (2009) The emergence of wearable computing. In J. McCann & D. Bryson (eds.) *Smart Clothes and Wearable Technology.* Oxford, Cambridge, New Delhi: Woodbridge Publishing Ltd. 3–24.

Macdonald, C. L. (2011) *Shadow Mothers: Nannies, Au Pairs, and the Micro-politics of Mothering.* Berkeley, Los Angeles, London: University of California Press.

Martin, B., & Wajcman, J. (2004) Markets, contingency, and preferences: contemporary managers' narrative identities. *The Sociological Review*, 52: 240–264.

McCann, J., & Bryson, D. (2009) Introduction. In J. McCann & D. Bryson (eds.) *Smart Clothes and Wearable Technology.* Oxford, Cambridge, New Delhi: Woodbridge Publishing Ltd. 1–2.

McDowell, L. (2008a) Thinking through work: Complex inequalities, constructions of difference and trans-national migrants. *Progress in Human Geography,* 32(4): 491–507.

McDowell, L. (2008b) The new economy, class, condescension and caring labour: changing formations of class and gender. *NORA,* 16(3): 150–165.

McGivern, G., Currie, G., Ferlie, E., Fidzgerald, L., & Waring, K. (2015) Hybrid manager-professionals' identity work: the maintenance and hybridization of medical professionalism in managerial contexts. *Public Administration,* 93(2): 412–432.

McMullin, J., & Dryburgh, H. (2011) Gender, age and work in the new economy. In J. McMullin (ed.) Age, *Gender, and Work: Small Information Technology Firms in the New Economy.* Vancouver, Toronto: UBC Press. 3–17.

Morel, N., Palier, B., & Palme, J. (2012) Beyond the welfare state as we knew it?. In N. Morel, B. Palier, & J. Palme (eds.) *Towards a Social Investment Welfare State? Ideas, Policies and Challenges.* Bristol, Chicago: The Policy Press. 1–32.

Nafus, D., & Sherman, J. (2014) This One Does Not Go Up to 11: The Quantified Self Movement as an Alternative Big Data Practice. *International Journal of Communication,* 8(2014): 1784–1794.

Naples, N. A. (2008) The challenges and possibilities of transnational feminist praxis. In S. Khagram & P. Levitt (eds.) *The Transnational Reader.* New York: Routledge. 514–526.

Neff, G. (2016) *Venture Labor: Work and the Burden of Risk in Innovative Industries.* Boston: MIT Press.

Neff, G., & Nafus, D. (2016) *Self-Tracking.* Boston: MIT Press.

Neff, G., Wissinger, E., & Zukin, S. (2005) Entrepreneurial labor among cultural producers: "cool" jobs in "hot" industries. *Social Semiotics,* 15(3): 307–334

Oakley, K. (2006) Include us out – economic development and social policy in the creative industries. *Cultural Trends,* 15/4: 255–273.

Oakley, K. (2011) In its own image: new Labour and the cultural workforce. *Cultural Trends,* 20(3-4): 281–289.

Ochse, R. (1990) *Before the Gates of Excellence. The Determination of Creative Genius.* Cambridge: Cambridge University Press.

OECD (2007) *Immigrant health workers in OECD countries in the broader context of highly skilled migration.* Paris: OECD. http://www.who.int/hrh/migra tion/2007_annual_report_international_migration.pdf   Retrieved   12 September 2015.

OECD Observer (2010) *International Migration of Health Workers. Improving international co-operation to address the global health workforce crisis.* Paris: OECD Policy Briefs. February 2010, 1–7.

Oudshoorn, N. (2012) How places matter: telecare technologies and the changing spatial dimensions of healthcare. *Social Studies of Science*, 42(1): 121–142.

Pagan, J. D., Cunningham, S. D., & Higgs, P. L. (2008) *Getting Creative in Healthcare: The contribution of creative activities to Australian healthcare*. Technical Report, Creative Industries. ARC Centre of Excellence for Creative Industries and Innovation. Queensland: Queensland University of Technology.

Poutanen, S., & Kovalainen, A. (2014) What is new in the 'new economy'? – Care as critical nexus challenging rigid conceptualizations. In J. Gruhlich & B. Riegraf (eds.) *Transnationale Räume und Geschlecht. (Transnational Spaces and Gender)*. German Sociological Association Book Series. Munster: Westfälisches Dampfboot Verlag. 176–192.

Poutanen, S., Kovalainen, A., & Jännäri, J. (2016) Construction of the female global top manager in The Economist. *International Journal of Media & Cultural Politics*, 12(2): 193–212.

Profita, H., Clawson, J., Gilliland, S., Zeagler, C., Starner, T., Budd, J., & Yi-Luenen, D. E. (2013) *Don't mind me touching my wrist: a case study of interacting with on-body technology in public*. Proceedings of ISWC'13, Sept. 9-12, 2013. Zurich, Switzerland.

Ramirez, E. (2013) How to download Fitbit data using Google spreadsheets: an update. At quantifiedself.com. Retrieved 15.7.2016.

Rosser, S. V. (1994) *Women's Health – Missing from U.S. Medicine*. Bloomington: Indiana University Press.

Schiebinger, L. (2000) Has feminism changed science?. *Signs*, 25(4): 1171–1175.

Sennett, R. (1998) *The Corrosion of Character. The Personal Consequences of Work in the New Capitalism*. New York, London: W.W. Norton & Company.

Sennett, R. (2006) *The Culture of the New Capitalism*. New Haven & London: Yale University Press.

Sennett, R. (2008) *Craftsmanship*. New Haven & London: Yale University Press.

Sleeman, C. (2016) Where do creatives cluster? *The Long+Short*, 1 Feb. 2016. http.thelongandshort.org/creativity/where-do-creatives-cluster. Retrieved 15.8.2016.

Stumpf, H. (1995) Scientific Creativity: A Short Overview. *Educational Psychology Review*, 7(3): 225-241.

Swan, M. (2012) Health 2050: the realization of personalized medicine through crowdsourcing, the quantified self, and the participatory biocitizen. *Journal of Personalized Medicine*, 2(3): 93–118.

Sweet, S., & Meiksins, P. (2012) *Changing Contours of Work. Jobs and Opportunities in the New Economy*. 2 ed. Los Angeles, London, New Delhi, Singapore, Washington DC: SAGE Publications.

Tao, H., & McRoy, S. (2015) Caring for and keeping the elderly in their homes. *Chinese Nursing Research*, 2(2–3): 31–34.

Tronto, J. C. (2011) Privatizing neo-colonialism: migrant domestic care workers, partial citizenship, and responsibility. In H.-M. Dahl, M. Keranen, & A. Kovalainen (eds.) *Europeanization, Care and Gender. Global Challenges.* Basingstoke: Palgrave. 165–181.

Topol, E. (2012) *The Creative Destruction of Medicine: How the Digital Revolution Will Create Better Health Care.* New York: Basic Books.

Tweedie, D. (2013) Making sense of insecurity: a defence of Richard Sennett's typology of work. *Work Employment Society,* 27(1): 94–104.

Vallas, S. (2011) *Work: A critique.* Boston: Polity Books.

Walby, S. (2002) Feminism in a global era. *Economy and Society,* 31(4): 533–557.

Walby, S. (2009) *Globalization and Inequalities: Complexity and Contested Modernities.* London, Thousand Oaks, CA, New Delhi: Sage Publications.

Wallerstein, I. (2014) Foreword. In W. A. Dunaway (ed.) *Gendered Commodity Chains: Seeing Women's Work and Households in Global Production:*: xi–xiii. Stanford: Stanford University Press.

Williams, C. C. (2011) Reconceptualizing men's and women's undeclared work: evidence from Europe. Gender, *Work & Organization,* 18(4): 415–437.

Williams, C. C., & Round, J. (2008) Re-theorising the nature of informal employment: some lessons from Ukraine. *International Sociology,* 23(3): 367–388.

Williams, C. L., Muller, C., & Kilanski, K. (2012) Gendered organizations in the new economy. *Gender & Society,* 26(4): 549–573.

Yeates, N. (2008) *Globalizing Care Economies and Migrant Workers.* Basingstoke: Palgrave Macmillan.

Yeates, N. (2012) Global care chains: a state-of-the-art review and future directions in care transnationalization research. *Global Networks,* 12(2): 135–54.

Zuboff, S. (1989) *In the Age of Smart Machine: The Future of Work and Power.* New York: Perseus Books.

www.carersuk.org (2016). Retrieved 13.8.2016.

CHAPTER 6

Envisioning the Future

6.1   Gender and Innovations: Turning the Tide

The scientific knowledge and science are profoundly part of the texture of the contemporary societies and embedded in the historical and contemporary developments in societies, as described in the previous chapters. Innovations build societies, economies and also everyday lives, and in the process they also change and transform our ways of living in many ways.

Innovations can be analysed as snapshots of a longitudinal process of events, or they can be singled out and analysed as individual inventions. This latter view is perhaps the most usual way of presenting innovations, especially when the history of innovations is narrated and reproduced in academic research. The focus in research most often is on individuals, businesses, organizations and their celebratory success stories. To take the whole chain of events, such as previous patents, practical developments and contextual features into account in every innovation description or case study is difficult, if not impossible. The drastic changes in any field can take place through major, ground-breaking innovations, but more often the major part of the innovations take place incrementally, building on previous work and developing in relation to previous innovations, patents and inventions. These two, largely differing ways of seeing innovations, either as radical or as incremental, also lead to very different outcomes when gender is being singled out in the analysis of innovations. When the focus

© The Author(s) 2017   169
S. Poutanen, A. Kovalainen, *Gender and Innovation in the New
Economy*, DOI 10.1057/978-1-137-52702-8_6

is on women only-type studies, women are easily seen either as heroes or as cogs in the machinery, whereas men are more readily presented as heroes and to a lesser extent as background figures in the innovation process.

Our analysis of gendered innovations in this book takes place at a time of increasing precariousness of economies based on innovations, growing digital economies based on platforms instead of stable organizational structures, widening the thinking about the innovations and gender. Some of the major, recent innovations are no longer restricted to corporation R&D or major laboratories only, but rather are developed through and with the help of networked vehicles such as start-ups, citizen science, and social innovations. The new forms of innovations, such as user innovations, crowd innovations, and democratizing innovations have expanded the innovation activities outside the corporation limits, into the public space and social space. For example, according to Von Hippel (2006), it is precisely the user-innovations that emerge from the heterogeneity of needs and from the limitations of traditional manufacturer-centered innovations, from social benefits of innovations, and, of course, from creativity. Von Hippel argues that intellectual property is useless to most innovators as only very large firms benefit from patents or copyrights. This is due to the great expense of patenting and renewal of patents, and in many cases benefits come at the cost of innovation. This is the case in defensive patenting and in the non-use of patenting due to strategic reasons in corporations.

The lack of women in the key fields of invention and innovation at large – science and technology included - is one factor in the scarcity of gender aspect in innovations. This is so even if women in science, for example, have been systematically researched, as we have shown earlier in the book, and even more so despite the growth of knowledge of how societies are missing the full innovation potential by not promoting women in science. Very often, the public eye is still on individuals and their exceptionalities, even if gendered innovations at large have become a viable research field. The persistency of gender differences and gendered segregation found both within the horizontal and at the hierarchical levels of innovation industries and in STEM education and careers are well documented and researched, but there is still much to learn. For example, the lack of policies that support work-family connections, and declining funding more generally, are persistently among the factors that stall women's participation in STEM fields (e.g. Bilimoria and Liang 2012), despite the growing amount of data on these matters. The practical ways of bringing gender knowledge into research institutions have however been established by for example, research-funding bodies in their gender main-streaming activities, as ways to tackle these questions more widely. The entrance of women into the science community and prevailing current minority in terms of positions held have also led to research which resembles the "history of great men", where women are placed as substitutes for men but the explanation for an individual's rise remains the same. As part of this explanation, the exceptionalities of women are emphasized. Even when raising the profile of women to the forefront, one of the major problems associated with this type of analysis is that it maintains the male norm as the measure of excellence and success, and categorizes women as exceptions. Feminist histories of science and the philosophy of science have systematically developed critical analyses since the 1970s and 1980s (e.g. Fox Keller 1985; Harding and Hintikka 1983; Nelson 1992, 1993), and have used gender as the focal point of understanding recent developments in science, technology, and related fields. The analyses of initiatives to increase the number of women in STEM careers emphasize that a portfolio of varied and multi-targeted gender equity initiatives is most effective in achieving gender equity transformations (e.g. Bilimoria and Liang 2014). The growing proportion of women in tertiary education is still not translating into a greater presence in research, as shown in UNESCO Science Report (Huyer 2015).

The definition of innovation we have adopted in this book is deliberately wide and does not only take into consideration the traditional definition of the commercialization of an invention through market demands. The big data development widens the concept of innovation as well. For example, the many ongoing scientific experiments where citizens are able to contribute by gathering data, are typical examples of web technology that can bring out new innovations (Neuphane 2015). Indeed, as markets are not a single phenomenon but several, and as markets can be both created and constructed, it seems that the distinction between inventions, innovations and imitations is not so clear-cut but instead is rather blurred. As we have shown through the case examples, innovations can start and also remain very local and restricted, and still be very successful.

All innovations are highly gendered in many ways. Research on gender and innovations has shown that workplaces are infused with different kinds of gendered practices, and that these practices take an active position in the ways that work – or innovation – becomes labeled as gendered. The phenomenon of gender and innovations visible in macro-level segregation figures has stalled. However, the macro-economic segregation figures do not tell the whole picture. Trends of innovation and invention activities reproduced across micro-level innovation activities have received significantly less attention in the analyses of gender, creative work and innovations, often largely due to the complexities of the processes (Poutanen and Kovalainen 2016).

With the economic and cultural globalization which transgresses national boundaries the questions of gender and innovation become even more complex: in addition to education and innovation systems, the location and place play increasingly important role in new ways for the innovations. The "end of place" has been proven to be unequivocally wrong, since creativity requires the cultural heterogeneity most often found in cities only. The cultural hybridization and cross-fertilization most often takes place where the dense and vibrant ecosystems of educational and cultural institutions for creativity to exist. In the USA and in UK the creative work is estimated to be concentrated geographically into broader metropolitan areas where creative class accounts for 30–35% of the workforce (e.g. Florida 2012). Yet, this estimate most probably undermines the number of those working with creative industries. Globally, the creative industries are growing in several countries remarkably, both in terms of employment and in terms of value added.

There is indeed a large body of literature and studies showing the relationship between cities, scientific, social and cultural innovations and economic growth, even if the exact relationship is difficult to pin down to individual factors. The fabric of urbanism varies, but often includes aspects of shared urban space, close proximity of services and knowledge creating and sharing institutions such as universities, start-ups and civic and cultural organizations. These types of vibrant hubs can be found everywhere in towns and cities with universities. How much do the platforms, cloud and gig economy increase the local and regional growth is worth exploring further. The technologies themselves are however not the solution or remedy as such, but they enable adaptations and make possible new innovations where gender emerges in new, yet unexplored ways in new settings.

6.2   Gender and Innovations: Widening the Field

The new economy is first and foremost the economy of existing and future digital platforms. Digital platforms are not the end but simply the means to do variety of activities, ranging from business to innovations and creative work. Gender is entangled in the digital platforms and new economy in myriad new ways, as we have shown in the book. The new economy also means digitalization. The historical growth of the industrial era, exemplified in the book by the birth of the electric washing machine, is a history of the gendered industrial production with specific gender divisions and layers of gendered work and jobs. The post-industrial, digital era of today and tomorrow leans on different types of divisions and gendered patterns. With the turmoil of digitalization, the platforms have become a powerful tool, not only as podiums for new types of jobs but also for companies and corporations as podiums for renewal. Where is gender located, and how is it subjected to new positions and possibilities in the discussion surrounding the new economy?

In the previous chapter we showed that despite the fact that innovations are becoming more and more crucial to contemporary economies, they are still highly gendered and also gender-biased. In addition to gender aspects of innovations, there are other elements in the contemporary economies that are worth the analysis. The gendered work and labor – whether local or global – challenge both the alleged disentanglement of skills/capabilities from personhood and the assumption that a 'corrosion of character' (Sennett 2006) will take place due to short-termism in the new economy.

The 'new economy' is said to have globally replaced the 'old economy' in recent decades. Although groundbreaking technological innovations that steer economy, society, transformation of the industrial sector and rise of the service sector are unquestionable developments, the reality is much more complex and entangled than is usually acknowledged. The multilayered relationships between the old and the new economies have been at the focus of the book. The complex entanglements and global significance of the rising platforms and gig economies and the changes they push forward have been discussed in the book at length. Much of the analysis on how the platform economies change the gender arrangements and societal gender order is yet to appear. It is still open where and by whom the innovations will be made in age of digitalization.

The complexity of the contemporary innovation fields – social and technological as well as material and immaterial – makes the gendering of the field very varied and multifaceted. They also bring in organizational change in many economies. As for much of the innovations, the short-term and long-term outcomes and the commercial breakthroughs are not self-evident or straightforward. The interconnectedness of some of the innovations and for many innovations the close path-dependency makes also the gendering processes differ from one innovation to another. Tackling this complexity requires meticulous and detailed analyses of the innovations and their birth processes. Further analysis of intersectionalities of gender, skills, capabilities and work related to innovations, is needed.

The platform economy has emerged in the footsteps of the technological development. The platform economy scales up and enables activities to grow in the digital world, disobeying the national borders and legislations. While the digital platforms may change the entry routes to the innovations, it remains to be seen whether the platforms enable gendering of digital innovations as such. What is evident is that the new technologies will continue to drive significant changes in the occupational skills needed through routinization of tasks and jobs, and through polarization of skills and capabilities. The platform economy will put new types of pressures for educational systems, science and research ecosystems included.

Indeed, most of the contemporary innovations, if 'social innovations' are excluded, are based on science and its applications in both adaptations in basic and applied research, as well as in technology. However, the fact is that relatively few contemporary studies focusing on the relationship between innovations and gender refer to the actual studies and classics of the field, where the relationship between gender and science have been explored. It is as if innovations and inventions are mainly understood in a contemporary and rather naively empiricist context that is unrelated to the broader questions of gendered science, technology and business. While the history of science, for example, reveals the gendered nature of many inventions and innovations, the actual and concrete processes in question are still largely untold. There is, for example, very little if any research on the gendering of financial values of innovation products and services.

Technology and innovations can bring equalizing elements into care, through the lowering cost structure. The future may be that with apps and personally gathered data through the algorithms of apps, medical expert knowledge that previously was owned by the professions such as medical doctors and highly professional groups gets transformed into user-driven, self-diagnosing gadgets and tracking devices. These devices require co-development, connectivity and user knowledge. The gender in design becomes crucial. The increasing consumer power requires simplified knowledge displayed, for example, as visually appealing and simplified charts, tables and measures. Research on 'doing gender' and 'gender and invention' concerning the adoption of innovations, diffusion and especially innovations in information communication technology will add new aspects to existing innovations studies. The adoption of citizen science in the data production brings in additionally the aspects of co-creation and co-development of innovations.

The future of health care will consist – to a growing extent – of highly complex, technology driven and even personalized medical care for those who can afford it. Typical for the future of the health care is that it is first and foremost technology-driven. Healthcare companies are currently reinventing both the diagnosis and the treatments of diseases, all with the help of the growing use of big data and technology driven appliances. Building algorithms that are good enough to accurately informing of any deviances in the data is the innovation that is needed most within the fields of the health care technology, irrespective of the type of care in question – be it intensive care unit estimating the care activities in the case of strokes or preventive care measuring the blood pressure. Any type of care is today full of technology, and its success is embedded in the usability of the technology. Thus, mastering technological devices and being able to be part of that development is the key to the future of care, also for migrant care workers, almost always women.

The gender gap in STEM fields is widely acknowledged. The statistics have established the striking gender division and low share of women in innovations: in the USA, women represent only 12% of the innovators, and that figure constitutes a smaller number than the share of women undergraduate degree holders in STEM fields, PhD students in STEM fields and women scientists in those fields (Nager et al. 2016). Even with this low share of women, the US innovation scene outperforms Europe. The low numbers of women among innovators in the patenting figures and in the innovation teams have been analysed widely both in the research of academic careers and in research dealing with innovations. The reasons for the low numbers have been elaborated and analysed in this book. The huge potential of competent and well-educated women is lost due to gender imbalance in the innovation activities, R&D and knowledge based invention work globally.

One of the main indicators for prospective innovators is STEM education, which in the USA is close to 90% of the innovators who majored in STEM fields at undergraduate level (Nager et al. 2016). This also demonstrates the importance of technologically driven and technology related innovations, despite growing interest in social innovations. Women's share in the STEM education has risen, but the strange disappearance of women from the innovation teams and groups raises several questions of the organizational culture and masculinity of the high tech company environment. Due to the low number of women the higher in the career path in STEM fields one looks, the positions where greater experience is required are highly gendered with only few women.

But gender is simply not just about women vs. men, and innovations are not simply about production of economic development and growth through technology and its new adaptations. The several attempts to define innovation range from Schumpeterian idea of doing things differently (Schumpeter 1939) to OECD definition in the Oslo Manual, where an innovation is defined as 'the implementation of a new or significantly improved product (goods or service), or process, a new marketing method, or a new organizational method in business practices, workplace organization or external relations' (OECD 2005: 46). While the distinction between invention and innovation may be quite clear, the direct use of the OECD definition of innovation may be highly problematic in research due to its normative nature.

The external environment has been recognized as having a strong influence on prospective innovations and entrepreneurship. But what exactly is the 'entrepreneurial climate' or 'entrepreneurial ecosystem' when it is analysed from gender perspective? The central theme of this book has been an exploration of the interconnectedness between innovations and gender and the ways in which these are in relation with each other in differing ways. Clearly, the contextual relationship – and assumed distance and closeness – between gender and innovations will undoubtedly affect the meanings and significance given to innovations in gendered analyses. Entrepreneurship is highly connected to innovations, especially in high tech and globally functioning start-ups and gazelle enterprises, where innovations are most often at the core of the business idea. Women are under-represented among the high-growth nascent entrepreneurs globally.

If economic austerity is one of the buzzwords of the twenty-first century development, so is also the platform economy, and increasingly so. As business model and as model for the economy to function the platform economy has indeed emerged in the footsteps of the technological development. The importance of platforms grows with the digitalization and with the growth of the IoT. The new business models with cloud-powered activities are growing, including new adaptations of social entrepreneurship and platforms as form of sharing economy. However, seeing cloud services as the natural platform for delivering new types of innovations is still developing even in the most advanced economies. Technology makes more technology available, and new innovations based on platforms are developing, as research on Amazon, Apple, Facebook and Google has shown. Work and working in platforms do not automatically relate to innovations, and do not automatically contribute to new innovations. The gig economy is yet to be analysed from gender perspective. Gig economy, for its part, allows short-term contracting, innovative ensembles and non-employer firms to flourish, but these do not happen without costs. The production of attractive and informative platforms requires new types of professional skills and capabilities. As mentioned earlier, vlogging through YouTube can be highly profitable, and vlogs are part of the creative work created and re-created through platforms, but they are also fully dependent on customers and their interests, and through that, the income from advertisement. Whether the gig economy can more generally offer the possibilities for new types of innovations, or is just an example of the future-type of specialized dead-end jobs, remains to be explored.

The ways in which platforms enable new innovations to become reality is changing the ways industries and services are arranged in economies. The disruption of traditional production and service models is currently taking place due to digitalization and digital, even global platforms. How do platforms function? Depending on the industry in question, platforms can take different forms. In many technological platforms the work turns into gigs and to different forms of self-employment where the work is no longer one but consists of many tasks and activities. Platform corporations use smartphones and digital technologies and there is seldom a limit in the ways these platforms function. Company such as Uber, for example, uses smartphones, and with its apps allows smartphone owners to hail a driver by using location software in the customer's phone and the driver's phone. Like most platform apps, the Uber app is connected to customer's credit account and majority of the payment goes to the driver who uses her/his own car and fuel. A minority fee goes to Uber. The innovative platforms use this same model and work for other types of services, from childcare to transfer of people and material goods.

The digital economies and platforms require highly educated workforce. The deepening collaboration between higher education institutions and the local and regional economies may provide several types of solutions for the pertinent questions such as business continuity and renewal, new business start-ups, entrepreneurial activities and entrepreneurial education, digital transformation of public and private sector, and the shortage of capable, skilled work force for the local and regional demand. In short, the innovation potential becomes connected to gender in different ways than previously.

How do we anticipate the integration of gender to take place in the larger innovation systems in the future years to come? The economic downturn during the first decennium of the twenty-first century did more or less globally introduce new types of pressures on and demands for the nationally bound innovation systems and their subsystems. These can be conceptualized as ecosystems for entrepreneurship and innovations that have national characteristics but are not restricted to work only within the nationally bound borders. Economic austerity and economic pressures nationally and supra-nationally do put forward several measures, such as cutbacks in the budgets that also change the innovation ecosystems for entrepreneurship. In the case of cutbacks to public funding, the innovation activities are geared towards the business sector. Global economic development has effects also on the ways gender becomes visible and present in different positions in the innovation and R&D activities in corporations. For these reasons, the widening of the topic of gender and innovations is needed, as the topic is not only restricted to patenting and science fields, but also covers widening societal and economic fields.

6.3   CONCLUDING REMARKS

Throughout the book we have explored gender, inventions and innovations, including the creative work leading to innovations. We have also argued that context for innovations is crucial because innovations seldom occur as a result of one person alone. In this book the new and old economy have been discussed in relation to innovations. The concept of "innovation" has been used since the sixteenth century. Originally the concept meant the idea of change, not creativity. In the late eighteenth and early nineteenth century, the forms of innovations were inventions in political and social organizations (such as laws, customs) and in technology. The intellectual understanding of innovations as contemporary immaterial and material commodities and foundation of the modern economies did develop through the early twentieth century. This was strongly influenced by the development of scientific breakthroughs and progress, ranging from nineteenth-century railways and factories to overall technological landscape that became to mean progress and innovation. In this understanding, innovation became to mean both progress and change. This has led us to raise the question: How can we explain the heavily gendered nature of innovations?

The disciplines differ in their understandings of innovation. The early sociology saw innovation as imitation, as activity and process of production of invention rather than contrasting these with each other. It was through early economics that dynamic change, the capitalism, was added into the equilibrium model in the form of the innovation as process of change. Over time, the research in economics was focusing solely on technological innovations and understanding innovation as the commercialization process of technological invention. This definition of innovation has become accepted as the reigning definition, of which the differences between open innovation, user innovation, or democratizing innovation are often understood as derivatives of technological innovations. More often than not, research takes technological innovation as a marker for general innovations and uses the term as generic sense.

It has been argued that intellectual property as ownership is not directly useful to most innovators as only large firms can draw benefits from large patents briefcase due to the high maintenance costs of patents. Only very large firms benefit from patents or copyrights due to their great expense, and in many cases even those benefits come at the cost of innovation (i.e. defensive patenting). As discussed earlier, innovations do not just happen, but most often they are part and parcel of the planned activities. The risks and expenditure on risk taking endeavors are part of the innovations, as in entrepreneurship. To analyse the complexity and the processes of innovations and their gendered nature entails often the understanding of entrepreneurial activities involved.

There is an extensive research on the global gender imbalance and gender gap in research and in academia, patenting, inventions and innovations included. Gender disparities in science, technology and engineering are enduring topics, and lack of women in STEM fields not only constitutes a gender gap but also signifies a very real loss of potential in terms of new ideas, inventions and innovations. Some key points of this discussion and the possible remedies, for example, in the form of newly established prizes for women innovators, have been brought up and discussed earlier in this book. Despite the long tradition of 'women and technology' research, the shift to analysis of the processes whereby technology and innovations are developed and used, as well as those whereby gender is constituted do not represent the mainstream focus in the innovation studies. The concept of technology, as well as innovations, is subject to historical change, and historically, for what we now think of technology or innovations, was something very different 20–30 years ago. A greater emphasis on women may easily be interpreted that women's innovations would be a niche that has remained the same or similar over the years. Part of gender research has continued to maintain the difference indirectly by focusing on 'other' types of innovations than technological ones, such as social innovations and user engagements.

The reasons for involving various stakeholders, businesses and civil society at large range from greater accountability to greater involvement and even agenda setting. The growth of the responsive modes of research and science systems are responses to this widespread request. The needs to accelerate and boost innovations in nationally important industries, services and technologies have become the key goals for most governments globally. How these ecosystems are connected, and in what ways they are gendered and have gendered effects, are among the pertinent questions for research on innovation and gender.

The studies focusing on gender differences and similarities in innovations emphasize the comparisons of men and women by numbers, and equality in terms of representations and numbers. Indeed, those studies that address gender in relation to number of innovations show that the number of women innovators is increasing, but that there are still fields and areas of inventions where gender analysis is non-existent despite the consumer type of the innovations, for example. But gender analysis on innovations goes theoretically much further than comparisons of mere numbers and figures.

The ways we understand innovations are not neutral. On the contrary, the mainstream understanding of innovation is stereotypically masculine and much of the innovation literature leaves out the feminine fields that are considered feminine, such as care. Innovations within care and social innovations remain invisible, because the emphasis on technological innovations is so overwhelming. These two fields – care and sociality – represent the gendered constructions of innovations. In care sector, which is highly labor intensive, the employee-driven innovations are at the core but defining the general innovation process for employee-driven innovations has proven difficult if technological process is not the focus of analysis.

The new economy and new platforms disrupt the earlier models for innovations and transform the business models. These will effect on gender and innovations: gaming and augmented reality are two aspects, cloud computing and services are another. We have asked what exactly changes in the development of new digital networked economy. Does that development offer new posts, positions and avenues for gender, or do we find remarkably well-known patterns of gendering emerging in the new economy? We have discussed the disruption and transformation of old and new economies through case examples that manifest the contrast between the old and the new economy. Gaming industry and tech-connected world offer new ways of including and focusing on gender, as shown through the cases. The focus on 'fixing the problem' by increasing the number of women in the coding world does not, however, solve the problem of gendered industry and corporations' networks. The multitude of ways the new innovations and services – with the growth of IIoT and IoT, for example – are gendered is not yet fully explored.

Throughout the book we have discussed the gendering processes of innovations, which do mean a very different and more dynamic gender aspect to innovations in comparison to gendered structures and gender differences. In a turbulent world it is ever more relevant to ask 'how does the gendering of inventions and innovations take place?' This question redirects the issue of gender and innovations towards the processes of innovations, and forces a re-evaluation of previous analyses of gender and innovations. Why and in what way have some products and services become gendered, and what is the gendering process of a particular product or service? What elements of design in high tech products such as wearable technology become recognized in the production as gendered and make the product gendered?

Innovation is a process that starts with an idea, but not all ideas lead to innovations. The point at which innovation turns into product, service, process or technology as commodity is an outcome of a complex process. Innovation, as an act of producing something new, novel or advanced, is about putting together complex pieces of immaterial and material resources. The commitment to actual work – as in innovation development – can be highly creative. This creativity and its entanglement with work widens the idea of innovations even more. In the book we have discussed the hybridization of work in relation to care work. This most often today includes also innovations and technologies, as well as creativity, thus breaking out from traditional view to innovations and inventions. The possibilities for new breakthroughs in an employee-driven corporate innovation culture, for example, are immense, as already shown in many corporations adopting such culture.

Innovation has intrinsic value of commodity: it is central to competitiveness and markets and growth of wealth. In this book we have discussed the patenting as one, albeit narrow, way to see the extent of activities and the types of activities that exist within innovations. The patents are markers of success and through them one can claim ownership to invention. Patenting is a seed for innovation, but it does not entail all aspects of innovations or their developments. In the previous chapters the research focusing on patenting and gender was valorized. More generally, this book has synthesized knowledge on the wider understanding of gender and innovations. Taking the global perspective, socially and societally even more important may be the social innovations and care innovations, that aim for increasing equality, human values and human dignity, for example. These types of innovations are too numerous to be named, and too varied to be dissected and to be analysed here but they are joined in being hugely important for the local and global developments overall. As the internet economy gains momentum, the economic landscape and its innovations are bound to change. The change concerns also science: it is becoming more mobile and networked. This will have huge implications to science-based innovations in the future. From the gender perspective, analysing the various paths to variety of innovations, the innovation processes and the end-products of the innovation processes, and recognizing those as innovations, are all crucial in the process of understanding the complexity of the question of gender and innovations in the future economies as well.

REFERENCES

Bilimoria, D. and Liang, X. (2012) *Gender Equity in Science and Engineering: Advancing Change in Higher Education.* New York: Routledge.

Bilimoria, D. and Liang, X. (2014) Effective Practices to increase women's participation, advancement and leadership in US academic STEM. In Bilimoria, D. and Lord, L. (eds.) *Women in STEM careers. International Perspectives on Increasing Workforce Participation, Advancement and Leadership.* Northampton, MA: Edward Elgar.

Florida, R. (2012) *The Rise of the Creative Class, Revisited.* New York: Basic Books.

Fox Keller, E. (1985) *Reflections on Gender and Science.* New Haven, CT: Yale University Press.

Harding, S., & Hintikka, M. (eds.) (1983) *Discovering Reality.* Dordrecht, Netherlands: D. Reidel.

Huyer, S. (2015) Is the Gender Gap Narrowing in Science and Engineering? In Schlegel, F., Schneegans, S., and Eröcal, D. (eds.) *Unesco Science Report. Towards 2030.* Paris: Unesco Publishing. 84–104.

Nager, A., Hart, D. M., Ezell, S., & Atkinson, R. D. (2016) The demographics of innovation in the United States. ITIF. Web-version: http://www2.itif.org/2016-demographics-of-innovation.pdf?_ga=1.194345133.1854841563. 1452803793. Retrieved 12.March 2016.

Nelson, L. (1992) *Who Knows? From Quine to Feminist Empiricism.* Philadelphia: Temple University Press.

Nelson, L. (1993) Epistemological communities. In L. Alcoff & E. Potter (eds.) *Feminist Epistemologies*. New York: Routledge.

Neuphane, B. (2015) A More Developmental Approach to Science. In Schlegel, F., Schneegans, S. and Eröcal, D. (eds.) *Unesco Science Report. Towards 2030*. Paris: Unesco Publishing. 6–8.

OECD (2005) *Oslo Manual: Guidelines for Collecting and Interpreting Innovation Data*. 3rd ed. Paris: OECD.

Poutanen, S., & Kovalainen, A. (2016) Professionalism and entrepreneurialism. In M. Dent, I. Lynn Bourgeault, J.-L. Denis, & E. Kuhlmann (eds.) *The Routledge Companion to the Professions and Professionalism*. London and New York: Routledge.

Schumpeter, J. A. (1939) *Business Cycles: A Theoretical, Historical and Statistical Analysis of the Capitalist Process*. McGraw Hill: New York.

Sennett, R. (2006) *The Culture of New Capitalism*. New Haven & London: Yale University Press.

Von Hippel, E. (2006) *Democratizing Innovation*. Cambridge, Mass: MIT Press.

INDEX

A

Abreu, Maria, 2, 21
Agency, 23, 112, 124, 159
Airbnb, 76, 140
Alexander Graham Bell, 10
  Graham Bell's telephone, 10
Algorithm, 53, 62, 69, 105, 111, 113, 157, 174
Amazon.com, 49
Anderson, Mary, 18
Android, 110, 123
Angry Birds, 122, 123
Apple
  app store, 47, 73
  GPS, 4
  Macintosh, 98
Apps, 76, 82, 104, 107, 111, 125, 157–159, 174, 177
Arnold, Frances, 35
Artisanal, 70, 79
Asia, 32, 118, 142
Asia-Pacific, 142
Atari, 98
Atypical working contracts, 51, 75
Augmented reality (AR), 105, 106, 107, 122, 125, 159
Australia, 31, 137
Authenticity, 77

B

Babbage, Charles, 111
B-Corporation, 80
Bendix, 14
Berners-Lee, Tim, 35
BlaBlaCars, 47
Black box, 34, 107
Blackstone, William, 11
  Blackstone washing machine, 11
Bose, 11, 14
Boston-NY area, 61
Branding, 4
  branding of innovations, 4
Breakthrough Prize, 36
  targets, 36
Broadband internet connectivity, 98
Brynjolfsson, E., 80, 83
Budget, 51, 83, 137, 143, 146, 178
Business
  activities, 78, 81, 97
  consultants, 85, 112
  gaming, 121
  model, 49, 52, 58, 71, 74, 76, 77, 83, 140, 146, 176
  opportunities, 51, 57, 77
  scalable, 62
  type of, 23
  venture-capital, 60, 62

© The Author(s) 2017
S. Poutanen, A. Kovalainen, *Gender and Innovation in the New Economy*, DOI 10.1057/978-1-137-52702-8

C
Canada, 31, 61
Capability, 55, 67
Care
    formal, 147, 156
    gendered, 149, 152, 153, 160
    global, 147, 150, 152, 154
    informal, 147, 154, 156
    medical, 175
    migrant, 152, 154, 158
    new economy, 149
    work, 147–154, 157, 151–154, 160
Car sharing, 17, 18, 177
Case, 3, 5, 6, 7, 11, 13, 15–17,
        27, 28, 35, 51, 52, 60, 62,
        64, 69, 73, 79–84, 109–119,
        125–126, 136, 143–145,
        146–154, 157, 169–171, 175,
        178–179
Change
    agent, 117
    changes in industrial structures, 51
    organizational, 173
    structural, 19
    technological, 57, 67, 136, 143
City, 76, 113, 119
Class
    creative, 71, 136, 137, 139, 143,
        145, 172
    social, 31, 135, 156
Co-creation, 175
Codeacademy, 109
Co-development, 175
Coding, 64, 108, 109, 111–112,
        114–116, 117–119, 125
Collaboration, 29, 31, 33, 50, 66,
        115, 145, 177
Commodore, 98
Company
    biotechnology, 29
    care, 155
    global, 114

machinery, 13
multinational company, 69
Competence, 102, 115
Competition, 20, 22, 23,
        28, 29, 34, 51, 68, 70,
        75, 100
Computer
    games (*see* Games)
    ubiquitous computing, 122
Conductive fibres, 160
Consumption
    digital games, 101, 104, 108
    game, 100, 124
    game-based learning, 103
    studies, 100
Cooking, 120, 153
Corporate femininity, 115
Corrosion of character, 48, 173
Couchsurfing, 77–78
Crafters, 80
Craftsmanship, 70, 136
Creativity
    class, 71, 136, 137, 139, 143,
        145, 172
    cultural and creative industries, 141
    dynamics of creative and innovative
        jobs, the, 59
    jobs, 137, 138, 139, 142
    occupations, 59, 135, 137, 142,
        145
    small creative firms, 59
    work, 1, 6, 7, 135–162,
        172, 177
Culture
    code, 77
    and creative industries, 141
    gendered organisations,
        organisation cultures and wider
        sets of institutions, 22
    gendered research, 22
    innovation, 9
Cylinder machine, 11

D

Dadification, 108
DARPA, 159
Decoded, 114
Design, 1, 50, 53, 55, 59, 69, 78,
    79–80, 99, 102–103, 105, 108,
    119, 122, 124–126, 136, 140,
    142, 145, 159, 181
  gender in, 175
Didi, 47
Diffusion, 174
Digital
  digitalisation, 54, 56, 57, 69, 73,
    74, 76, 85, 97, 105, 107, 122,
    136, 159, 173, 176
  games, 101, 104, 108
  new economy, 97
  platforms, 47, 52, 54, 73, 75, 79, 85,
    97, 107, 121, 140, 172, 174
  tailoring, 159
Disruptive force, 51
Diversity, 27, 54, 56, 100, 104, 117,
    118
Doing gender, 174
Dot-coms, 47
Draining closet, 15
Dynamics of creative and innovative
    jobs, the, 59

E

Ebay, 83
Economic activity, 23
Economic growth, 53, 55, 56, 62, 172
  and innovations, 172
Economist Innovation Award, 36
Economy
  dot-com, 47
  economic shocks, 51
  gig, 51, 52, 73–75, 76, 78, 85, 140,
    141, 172, 177
  interconnected nature of the, 70

new economy, The, 38, 47–56, 58,
    63–70, 75–78, 80, 83–85,
    97–127, 135–139, 143,
    146–147, 149, 152, 153–154,
    160–161, 172
  old economy, the, 2, 37, 80,
    152–154, 160, 173
  peer, 76
  platform, 4, 6, 47–85, 98, 103, 122,
    138, 140, 173, 176
  sharing, 73–74, 76–78, 83, 85
  world, 70, 150
Ecosystem
  innovation ecosystem, 3, 9, 24, 178
  R&D ecosystem, 3
Edgell, R. A., 54, 75
Edison, Thomas, 10, 20
Education
  educational segregation, 24, 31
  policy, 6, 32
  segregated patterns, 25
Ejermo, Olof, 5, 6, 21, 24
Electric
  light, 1, 10
  vacuum cleaner, 13
Embeddedness, 64
Embodiment, 64, 68
Employee, 85, 113, 141, 180
  -led innovation, 98
Employment
  crowd, 52, 73
  future, 31
Empower, 114
Enactments-perspective, 57
Engineering
  software, 68
  women in, 31
Entrepreneur
  academic, 65
  female, 29, 60
  inventor-entrepreneur, 20
  male, 60

Entrepreneurial
  activities, 57, 161, 177, 179
  labor, 138
  professions, 85
  projects, 60
  renewal, 75
Entrepreneurship
  academic, 65
  and gender, 3
  innovation, 54, 109
  small-scale, 78, 140
Etsy, 78–80, 140
  crafts and design, 78
Etzkowitz, H., 2, 31, 33, 34, 50
Europe, 15, 20, 26, 30, 32, 33, 52, 53,
    61, 62, 73, 76, 80, 119, 150,
    152, 154, 175
  code week, 118
European Commission, The, 29, 118
Everyday objects, 120, 122
Expert, 64, 85, 119, 124, 150, 158,
    174
  jobs, 85, 119
Expertise, 64, 80
  professional, 64
E&Y Barometer, 62

F
Facebook, 49, 81, 105, 176
Fair play, 103
Family influence, 31
Female
  entrepreneurs, 29, 60
  innovators, 2
Feminine
  corporate femininity, 55, 116
  corporate feminism, 116
Femininity, 23, 116, 147
Feminism
  corporate, 55, 116
  feminist research, 99

Filament lamp, 10
Finland, 15, 16, 17, 35, 54, 76,
    110, 146
Fitness wristband, 144
Flexibility, 73, 116
Florida, Richard, 135–138, 143
FujiFilm
  gendering of corporation, 82
  transformation, 82–83
Funding
  government, 62
  instruments, 61
  public, 72, 177
  research, 23, 33, 34
  venture capital, 60
Future
  employment, 31
  labour, 75, 162

G
Game, 99
  consumption, 100, 124
  girls', 99
  social network, 98, 102
  studies, 101, 109
  virtual on-line, 99
Gascoigne, Adriana, 113
Gebhard, Maiju, 16
Gender
  analysis in healthcare, 148
  composition, 55
  differences, 2, 5, 31, 60, 64, 65,
      100, 101, 104, 108, 170, 180
  disparities, 2, 19, 27, 30
  effects, 33, 154
  gap, the, 2, 3, 21–24, 26, 30, 60,
      62, 108, 112, 118, 175
  gap in technology, 108
  imbalance, 2, 19, 24–25, 61, 115, 175
  and innovations, 28, 64, 83, 98,
      169, 171, 172, 180

labelling, 19
as process perspective, 48, 55, 176, 177, 181
and venture capital, 62
Gendered
disparities, 19, 25, 26, 31
division of household work, 13
organisations, organisation cultures and wider sets of institutions, 22
promotion patterns and practices, 22
research cultures, 22
Gendering
definition, 97, 101
innovations, 14, 170
Geography, 97
Ghiasi, Gita, 19, 31
Gig economy
as form of economic activity, 85
as professional work, 74, 141, 177
Gigle, 146
Girls in ICT day, 114, 118
Girls in tech initiative, the, 113
Girls-in-technology, 114
Girls who code, 108
GlaxoSmithKline plc (GSK), 29
Global corporations, 53, 117, 153
Global economic competition, 51
Globalisation
diversity of, 54
1.0, 53
2.0, 53
3.0, 53, 58
X.0, 58
Google, 36, 47, 49, 62, 69, 73, 110, 118, 176
app store, 47, 73
Grand narrative, 18
Granovetter, 64
Grinevitch, 2
Gutierrez, B., 103

H
Handicraft design, 79
Healthcare
gender analysis in, 148
services, 149
Hello Ruby, 109, 110
History of innovation, 18, 20, 169
Hitman: agent, 103
Household
appliance, 11, 12, 13
gendered division of household work, 13
logic, 13, 14
Hub, 58, 71, 172
Human capital, 3, 23, 69, 85, 151, 161

I
IBM PC, 98
ICT
sector, 118
women in ICT, 118
Idea, 5, 18, 21, 22, 37, 52–54, 60, 62, 66, 70, 75, 78, 81, 85, 99, 109–110, 115–117, 123, 125, 135–138, 146, 147, 153, 160–162, 176, 178, 181
scalability of ideas, 9
Identity
occupational, 162
work, 162
Immaterial
effects, 84
service, 1, 5, 49
India, 50, 52, 54, 61, 118
Industry
changes in industrial structures, 51
industrial production, 14, 15, 16, 47, 71, 173

Information communication
    technology
    innovations in, 174
Ingress, 105
Innovation
    Adoption, 174
    branding of, 4
    climate, 5
    contemporary system, 20
    culture, 9
    dynamics of creative and innovative
        jobs, the, 59
    as economic drivers, 47
    ecosystem, 3, 9, 24, 178
    entrepreneurship, 54, 108
    and gender, 1, 6, 180
    gendered nature of, 2
    history of, 18, 20, 169
    incremental process of, 2
    interconnected innovations, 1
    policy, 2, 32, 56, 62, 72
    process, 3, 4, 6, 36, 38, 170,
        180, 182
    revolutionary, 10
    social, 6, 34, 56, 67, 68,
        110–113, 170, 174, 175,
        180–182
    society, 3
    technological, 1, 14, 20, 35, 48, 53,
        63, 67, 120, 147–154, 173,
        179, 180
    women innovators, 20, 28, 29, 179,
        180; women Innovators [prize]
        winner, 28
    women's, 179
    workability of, 33
Instagram
    as new economy, 80
    as platform economy, 81, 82
International patent classification
    (IPC), 27
    IPC codes, 27

Internet of things, the, 120, 122, 159
    Industrial IoT, 122
Invention
    gender and, 174
    women's, 1
Inventor, 4–6, 11, 20–21, 23–24, 27,
    35, 66, 70
    entrepreneur, 20
iPhone, 4, 120
Israel, 61, 102

J
Jobs, Steve, 115
Jung, Taehyun, 5, 6, 21, 24

K
Kenney, Martin, 50, 74, 75, 76, 136
King, James, 11
Knowledge
    burden of, 5, 6
    creation of, 172
    global, 77
    interdependence of, 58
    service clusters, 51
Kodak
    girl, 81–82
    as old economy, 80

L
Labour
    entrepreneurial, 138
    future, 75, 162
    markets, 11, 24, 34, 47, 51, 64, 73,
        74–75, 85, 119, 139, 151, 152,
        161–162
    national, 51, 151
    venture, 140, 161, 162
Lady geek, 112–113

Lara Croft: Tomb Raider, 103
Leaky pipeline, 2, 19, 22–23,
    26, 69
Let the Girls Learn, 118
Lingering effects, 70
Little miss geek, the, 112–113, 115
Liukas, Linda, 109–113
Local living experience, 77
Lovelace, Ada, 109, 111–112, 113

M
Machine
    cylinder, 11
    metaphor, 9
    washing, 10–15, 121, 173
Made with Code Initiative, 118
Maiju Gebhard, 16
Market
    consumer, 12, 82, 127, 142
    global, 68, 77, 79, 80, 122, 127,
        142, 158
    labour, 11, 24, 34, 47, 51, 64, 73,
        74–75, 85, 119, 139, 151–152,
        161–162
    logic, 13, 14
Marketing, 4, 82, 104, 141,
    142, 176
Mary Anderson, 18
Masculine
    'do masculinity', 68
    Hegemonic masculinity, 23
    hyper-masculinty, 108
    masculinity, 15, 23, 31, 68, 100,
        102, 107, 115, 175
Mathilda Effect, 33
Matthew Effect, 33
Max Payne, 103
Maytag, 13, 14
McAfee, Andrew, 83
Mechanism
    funding, 34, 72

reward, 10
support, 32
Media
    computer, 107
    landscape, 100
    social, 49, 82, 113
Mentoring, 32, 117
    programme, 32
Merton, R. K., 33
Migrant, 70, 139, 151, 152, 154, 175
    migration, 150–152, 161
Millennium Technology Prize, 35, 36
Minkoff, Rebecca, 125
Minority groups, 3, 119
MIT area, 61
Mixed-methods design, 99
Mobile phone, 1, 4, 10, 105, 126
Mobility, 22, 150–152
Moghadam, 65
Mortal Combat, 103
Mozilla Firefox, 110
Multiplying effect, 70

N
Nählinder, J., 19, 24
Narrative, 18, 109, 115
New forms of work, 52, 54, 73, 75,
    122
Nikki Kaufman, 125
Normal (company), 126

O
Old forms of work, 76
Opportunity, 29, 33, 65, 78, 79, 104
    business, 51, 57, 77
Organic urban vibrancy, 77
Organisation
    cultures, 22
    gendered, 22

Organisation (*cont.*)
   non-profit, 108, 118
   political and social, 178
   profit, 116
Organizing, 73, 117, 122
Outreach, 32
Outsourcing, 50, 52, 53

P
Parsons, Kathryn, 114
Patent
   co-patent, 32
   International patent classification
      (IPC), 27
   patenting; female activities, 26;
      patterns, 21, 24
   patent litigation, 20
   smartphone patent wars, 20
   women's patents, 26–28, 32
PC, 49
   IBM PC, 98
Personal choices, 30
Pinterest, 82
PISA Study, 101
Platform economy
   business models, 58, 74, 76, 83,
      140–141, 146, 176
   as form of economic activity, 47–86
   as professional work, 138
Platforms
   digital, 47, 52, 54, 73, 75, 79, 80, 85,
      97, 107, 121, 140, 172, 174
   economy, 4, 47–86, 103, 127,
      174, 176
   informative, 141, 177
   mobile, 138
   technological, 75, 120, 143, 147,
      149, 177
   See also Gig economy; Sharing
      economy

Pokémon Go, 105, 106, 122–123
   PokéStops, 124
Polarization of the work force, 51, 75
Policy
   analysis, 32
   education, 6, 32
Popular Culture, 103
Poutanen, Seppo, 2, 23, 31, 32, 37,
      38, 55, 66, 72, 85, 107–108,
      136, 153, 172
Power, M., 58
Practice-perspective, 57
Prince of Persia, 103
Printer
   Ink-jet, 49
   laser, 49
   3D, 125–126
Product, 1–6, 11–14, 47, 55,
      67, 71, 73, 81, 83,
      121–122, 126, 144–145,
      176, 181
Productivity
   paradox, the, 57
   total factor, 83
Programming, 109–112, 114–116,
      118
Prototype, 11, 18

Q
Qualifications, 52, 54, 64, 71, 75, 84,
      117, 139, 141, 152
Quirky, 125

R
Race, 97, 102
Radical innovation, *see* Innovation
Rails girls, 110–112
Ranga, Marina, 32–34

Reality
    augmented reality (AR), 105–107,
        122, 125, 159
    mixed reality (MR), 105, 106
    mobile augmented reality, 105
Refrigerator
    internet, 121
    smart, 121
Region, 58–60, 62,
        63, 142
Research
    career, 22, 24, 30
    gendered cultures, 22
    landscape, 51, 72
Research & Development (R & D), 3,
        34, 54, 55, 57, 60, 170
    R&D and gender, 7
    R&D Ecosystems, 3
Reshma Saujani, 108
Resource, 3, 23, 30, 33, 66,
        72, 78, 83, 115, 145,
        161, 181
    'untapped resource', 30, 115
Ridgeway, 65
Robotization, 54, 75,
        108, 113
Rouvinen, Petri, 75
Rovio, 123
Rubery, Jill, 73, 76
Ruby Rails, 110

S
Saarinen, Karri, 110
Satoshi Tajiri, 106
Saxenian, AnnaLee, 71
Scalable
    business, 62
    scalability of ideas, 9
Schiebinger, L., 2, 19, 22–23, 38,
        65, 148
Schmidt, B., 25, 65

Scholastic performance, 101
Science
    policy, 30
    prize, 37
    women in science award, 37
Segregation
    educational, 24, 31
    education patterns, 25
Sennett, Richard, 48, 54, 70,
        75, 76, 136, 161,
        162, 173
Seppälä, Timo, 73, 74
Service, personal, 148, 150
Sexuality, 100, 102
Shanghai, 71
Sharing economy
    business models of, 77, 80, 83
    as form of, 176
    See also Gig economy; Platform
        economy
SHE figures, 30
Shopping, 120, 125, 155
Siemens, von, 84
Silicon Valley (Bay Area), 61, 62
Skills
    ICT, 115
    language, 152
    mathematical, 125
    occupational, 162, 174
    professional, 74, 141, 177
    technical, 110, 112
Small-scale production, 70
Smart home appliance, 120
Smith-Doerr, L., 2, 3, 25, 27, 30
Snapchat, 82
Social
    Entrepreneurship, 176
    innovation, 6, 34, 56, 67, 68, 110,
        111, 113, 170, 174, 175,
        180, 182
    network, 23, 61, 98, 99, 102, 115,
        123, 161

Software, 49, 54, 68, 69, 112, 141, 146, 153, 177
software engineering, 68
Solow, Robert, 57
Soviet Union, 15
Stakeholder, 9, 28, 180
STAM, 6
Start-ups, 2, 49, 50, 55, 58–61, 72, 98, 114, 125, 126, 146, 161, 170, 172, 176, 177
STEM
    education, 22, 138, 170, 175
    fields, 6, 22, 27, 64, 66, 109, 113, 118, 175, 179
Sugimoto, C. R., 25–27, 30, 33
Super Mario Bros, 103
Support, 32, 36, 58, 63, 66, 72, 118, 151, 155, 160
    mechanism, 32
Sweden
    Northern Sweden, 76
    Swedish Medical Research Council, 33

T
TaskRabbit, 47, 76, 140
Technical, 1, 13, 23, 52, 56, 60, 63–65, 82–83, 100, 104, 108, 110–112, 120–121, 147, 157, 159
    background, 23
Technology
    bubble, 47
    development, 4, 12, 14, 47–49, 51, 52, 58, 66, 74, 126, 146, 149, 174, 176
    drivers and changes in industrial structures, 51
    impact, 27

innovations, 1, 14, 21, 35, 48, 53, 63, 66, 120–121, 147–157, 173, 178, 179
    university transfer, 9
    wearable, 113, 125–126, 155, 157, 159–160, 181
Telemedicine
    algorithms, 157
    connected, 158
    remote diagnostics, 158
Thompson Reuters, 19
Thor
    gendering, 12
    washing machine, 12, 13
Time, 103
Torvalds, Linus, 35
Transformation of skills, 51, 75
Transnationalism, 152, 161
Turkle, Sherry, 107
Twitter, 49

U
Uber, 47, 52, 73, 76–78, 140, 177
Underperformance, 26
Underrepresentation, 2, 25, 27, 30
Unemployment
    cyclical, 84
    frictional, 84
    higher levels of, 51, 75
    structural, 84
Unesco, 119
University
    careers, 66
    cultures, 26
    education, 22
    technology transfer, 9, 28
Upwork, 47, 73, 78
U.S. Bureau of Labor, 119
User
    -experience, 5
    -friendliness, 5

V

Vacuum cleaner, 13
  electric, 13
Vallas, Steven, 54, 75, 76, 139
Value capture, 74
Value creation, 4, 28, 50, 52, 54,
  69, 73
  product, 4
Venture capital
  gender and, 62
  sector, the, 60
Venture labor, 161
Vine, 82
Vlogging, 141, 177

W

Wages, 54, 147, 153
Wajcman, Judy, 162
Washing machine, 10–15,
  121–122, 173
Wellbeing, 72, 148–150, 159
Western, 21, 52, 53, 55, 69,
  148–150
Whittington, K. B., 4, 23, 25,
  27, 30
Wikipedia, 49
Women
  educated, 175
  in engineering, 31
  European, 29
  innovation, 178

innovator, 20, 29, 179, 180;
  Women Innovators [prize]
  winner, 29
invention, 1
  patents, 27, 33
  professional, 113
  in science award, 37
  Start-up Lab, 114, 116
Work
  care, 147–148, 147–154, 160, 174
  creative, 1, 6, 7, 135–162, 172,
    172, 177
  gendered, 22, 48, 51, 54, 149, 152,
    173
  household, 11–16, 56
  innovation, 33, 57, 66
  knowledge, 63–65, 124
  media, 140
  skilled, 68, 138, 177
  voluntary, 116
Workability of innovations, 33
Work-family conflict, 34
Workplace, 27, 31, 32, 66, 70, 135,
  140, 161, 171, 176

Z

Zipcar, 78
  car sharing, 78
Zuckerberg, Mark, 36, 113, 115
Zysman, John, 48, 50, 54, 74, 75, 76

Printed by Printforce, the Netherlands